Francis Galton: Selected Papers, Notebooks, & Diaries

Francis Galton

Selected Papers, Notebooks, & Diaries

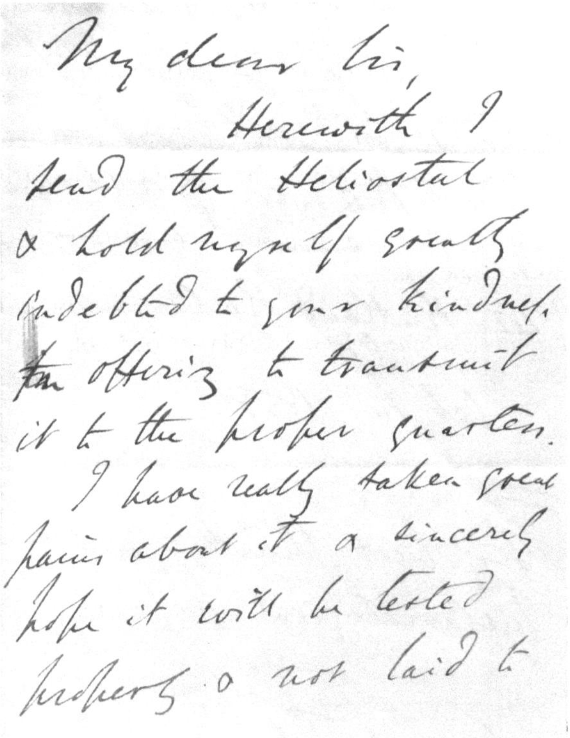
My dear Sir,
Herewith I send the Heliostat & hold myself greatly indebted to your kindness for offering to transmit it to the proper quarters.
I have really taken great pains about it & sincerely hope it will be tested properly & not laid to

Edited by Gavan Tredoux

galton.org

Francis Galton: a Lifetime of Exploration.

1. *Francis Galton's Nature and Nurture:* 1822–1865.
2. *Francis Galton's Genius:* 1865–1911.

Supplements:

1. *Sir Francis Sacheverel Darwin:* Charles Darwin's Adventurous Uncle.
2. *Francis Galton at Cambridge:* Letters and Diaries, 1840–1844.
3. *The Diary of Charles John Andersson:* 1850–1851.
4. *Francis Galton on Mars:* the Discontinuous Variation Notebook.
5. *Francis Galton and Alphonse de Candolle:* Notes and Correspondence.
6. *Francis Galton and Alfred Binet:* Correspondence.
7. *Francis Galton:* Selected Papers, Notebooks and Diaries.

Figure 1: Francis Galton in 1907 at age 85.

Contents

Preface

This volume collects important diaries, reviews and critical editions of selected memoirs by Francis Galton.

The material was originally prepared for the biography *Francis Galton: a Lifetime of Exploration*, to be included as appendices to the two main volumes, *Francis Galton's Nature and Nurture: 1822–1865* and *Francis Galton's Genius: 1865–1911*. But the material grew so bulky that it has been placed in a separate volume, one of seven supplements containing primary source material.

The papers in particular are not meant to be representative. For example, the technical memoirs on regression and correlation have not been included—merely because no critical editions of those were needed for preparation of the biography.

All material is provided for educational purposes only.

Gavan Tredoux
October 2022.

Chapter 1

Biographical Notes on the Galton Family

[Galton Papers, UCL, GALTON/1/1/10/1.]

f.2.r

Galton. This family is curious, chiefly through the diversity of its hereditary elements. A love of tabulation is instinctive to many of its members; I have alluded to this in my remarks on innate tastes.

First generation. Samuel Galton FRS & one of the early members of the Linnaean Society (memoirs on colour in the Phil Trans, & author of a once popular little book on birds). A wealthy Quaker whose disposition was full of contrasts. At the close of the century his house was one of the social centres of the scientific & literary society of the Midland counties, which then included men of note as social reformers, free thinkers & chiefs of new industries. (See description of him in the memoirs of his daughter, Mary Anne Schimmelpenninck, small edition of 1858, I 250).[1] His wife had strong will and intelligence, and a peculiar stateliness of manner; combined with rather a pedantic turn of mind & many singularities. She was descended, through a very mixed line maternally, from Barclay the Quaker "Apologist" and was sister to Capt Barclay of Ury, a man of numerous adventures, one of the strongest, perhaps the very strongest person, in England, the pedestrian of 1000 miles in 1000 hours, noted in coaching & hunting an-

[1] Mary Anne Schimmelpenninck née Galton (1778–1856). Sometimes given as Marianne.

nals, and in later years as an agricultural reformer (see his memoirs[2]).

Second generation 4 males and 3 females; 1 at least of the males, & 1 of the females deserve mention. (1) Theodore; Reputed to have had many gifts, as well as the instincts and aptitudes of a traveller. He died of plague on returning from the East while still young. See a most laudatory and affectionate memoir of him. I know not by whom, in the Annual Register 1810 p. 385[3] ~~(?) Conspicuous for his love of society, delight in rapid travel & power of enduring fatigue, combined with energetic organising aptitudes which he showed when at home.~~

(1) Lucy, [sic, Marianne] married Mr. Schimmelpenninck, authoress of Memoirs of Port Royal &c. I have lately heard no less than 3 notable persons who knew her in her mature and declining years, when she had become a pious Moravian, speak of her influence for good over her acquaintance

f.2.v

Notes from Notable Families.

f.3

in the highest terms; one of them, who has lately received high honor from the State, using expressions almost of adulation.

Third Generation; 7 sons & 7 daughters, of whom only 2 fall technically under notice, though a full description of the family tendencies would shew much that was instructive from an hereditary point of view.

(1) Captain Douglas Galton FRS, formerly RE, Secretary of the Railway Department then of the financial department of the War Office & now Director of Public Works, distinguished by extreme official activity, in all these capacities, especially in connection with the sanitary improvement of hospitals & barracks, & economy of fuel. (Descended maternally from the Strutt family, noted for the mechanical aptitudes of its founder & of 3 of 4 of his numerous descendants)

(2) myself, for reasons given, under "Darwin".

---

2 Not located, despite intensive searching. This appears to be a mistake.

3 Anonymous 1810.

f.4

The founder of the good fortunes of the Galton family was my great grandfather John Samuel, born in 1720. He was a Somersetshire Quaker, who migrated to Birmingham and there manufactured muskets which were bought by the armies of England and the Continent and were even exported for the African slave trade. Of his parentage I only know that his father began life with £10,000 which he neither increased nor diminished and that John Samuel had 5 brothers or sisters who all died unmarried or at least childless and of whom the only one who has figured conspicuously in the annals of the family was John;—the Uncle John of later Duddeston days. Those who may be curious to know more should dive into the registers of Yatton, Somersetshire, which is the earliest home to which the family has been traced.

John Samuel founded our fortunes but the individuality and dispositions of his descendants are mainly due to the peculiar character of his only surviving son's wife; Lucy, sister of the famous Quaker pedestrian, pugilist and trainer—Captn Barclay of Ury, the lineal representative of the old "Apologist".

We will now endeavour to portray the three individuals of whom I have spoken, namely John Samuel the father, Samuel his son & only representative & Lucy, Samuel's wife. To give fixity to our ideas, let us select some particular year, say 1795, when the father had reached the goodly age of 75, and had succeeded in his wordly enterprises & handed the furtherance of them to his son. When the son was 42 years old, & married 18 years & had a growing family of staid ~~Quaker~~ children & when they all lived in the beautiful suburban residence of Duddeston. I believe a quarter combination of Quaker simplicity & narrowness, commercial skill, with encyclopedic education, artificial mannerism & profuse expenditure.

f.5

As John Samuel had 5 brothers & sisters none of whom left issue so his eldest son Samuel had 6 brothers & sisters who died childless before the days of my informants. The family was therefore condensed at one time into old shrewd John Samuel the wealthy & retired manufacturer, Samuel his son who carried on his father's business unwillingly for he had wider tastes, & Samuel's wife, my grandmother, who possessed one of the quaintest combinations of qualities that ever composed a truly harmonious character. Besides these near the close of the last century were children & babies, my future Father, Uncles & Aunts, of whom much more hereafter, and also "Uncle John".

It is not possible for me to give a just analysis of John Samuel's character which now only lives in the recollection of those who were children when they knew him, but his letters shew that he was more than usually educated for his class, his age and his town. He was married early to the daughter of his employer, he received a good position with the lady and was taken into partnership with the father. He was a stiff Quaker but above all things a hard man of business. Thus it fell out that the Birmingham Quakers grew scandalised at the vast business connected with war being carried on by their prominent co-religionist. John Samuel maintained their objections came too late, for his gun trade had grown under their eyes without a word of remonstrance, that it was now established & more lucrative than he cared to abandon. The Quakers regretted they had not thought of warning him before but adhered to their views and made him to understand he must either give up the trade or suffer dismissal from their Society. I am sorry to say, he compromised the difficulty in a wordly manner. He retired nominally from the business but made his son carry it on; the standard was diminished and the Quakers said no more.

f.6

combination of contradictory qualities existed nowhere than at Duddeston. The family were rigid Quakers and aimed at simplicity and seriousness, on the other hand they were indoctrinated by the schools of Madame de Gentis and of the French Encyclopédists until they were utterly artificialised. Every action was *calculated* for culture and for science, but such ungenial culture & such narrow science! Staidness was mistaken for dignity, and measured words for poetry. ~~There was an air of homilies and moral phrases~~ The children were never children at Duddeston but the old dogs & horses were buried under stuccoed urns ~~of a would be classical shape in~~.

f.7

from the business but assigned to his son in his own behoof. The Quakers ceased to object. Poor Samuel who had business was [gently] broken in to its duties his schoolboy half holidays his father assigned to cyphering in the counting house He was generous but never lavish for he overestimated the value of money. His grandchildren thought him kind but he was so hard & unsympathetic a father that his son would rarely speak of his early days.

f.8.r: Aunt Brewin, Sophie Galton

All grandfathers better liked by grandchildren than children. GtGrandfather[4] was harsher to his children than Grandfather. GtG shrewd—very courteous—& well-educated for those days. Grand very clever so Gmother rather in advance of age in which they lived for their class—nonsense about the King's daughter. GrGrMother died 3 days after birth of Grandmother—no register found of her birth. GGFather was a Brauvist, GGM a catholic & therefore none probably cared to register the birth in the Quaker's meeting. She always lived with her maternal grandfather, apprenticed to a house in Cheapside a corner—excellent place for King to shew himself. His master had received 3 Georges in succession. Duke York bounded about room & picked up child—curious coincidence. King married Birm Quaker.

Aunt B.[5] never heard Grandfather speak of his family whence she concludes they were no great things. Her mother said <u>her</u> GGFather (believed to live near Taunton) had £10,000 never increased or diminished it. Uncle John was one of 7 or 8 children, a merchant. AB has seen his merchant's books. Her GFather was the only one who had children. "Uncle John" had High Ham manor whether bought or not AB does not [know].

Father was much with Dr Priestley & though he never said so AB thinks he was a Unitarian.

MSS

[4] All relationships in these sections based on Aunt Brewin's information are relative to her, not to Francis Galton. Thus F is Samuel Galton jun., GF is Samuel Galton Sr, GGF is John Galton.

[5] Aunt Sophie Brewin née Galton (1782–1863). 'AB' below. It is not clear when these notes were made, or whether they were based on diaries or on interviews.

AB G.Father loved business & made her Father spend his 2 half holidays in counting up accounts in his counting house to his gt disgust. He was sent to the academy at Warrington. Priestley was a tutor[6]—Miss Barbauld, Mr Aikin were there. His earlier days were very unhappy. He would not talk of them. The early Duddeston days were very unhappy. [M de Gentis ...] AB mother tried to educate herself

f.9: Aunt Brewin, Sophie Galton

to educate Mrs Schim—became ill & the charge of the children went to other hands.

Mother had not the slightest idea of the management of children. She fancied ridicule was a good way of managing them. She established a "medley" contributed by all & read on Sunday's evening. She took that opportunity of writing to ridicule them. ~~then AB after whooping cought had always a cough to clear her throat for years—others said she could help it, she could not~~ Mother carried a book in her pocket to note down everything—a horror to the children. The children copied the ridicule—bad habit. The papers said Father too much immersed in business to care much her & mother [suppressed] all the conversation—there was no general information & the children all felt it. He was employed at something or other all his working hours & managed to employ ½ a dozen people besides.

AB GM brought up with wealthy people—would have been spoilt by adoration had she not been [...]

G made money more by driblets & liked money & though generous not lavish. This probably influenced GF will, which nearly drove F mad. He was ill & mad subsequently owing to Mrs S. The Quakers did not approve of GF & F making firearms & threatened dismissal at length, there was correspondence. GF did not like to abandon the business or be abandoned by the Quakers so he nominally gave it over to [F] who hated it & would not let him throw it over under threat of disinheriting.

GF thought M very extravagant because she spent money in the way she had been used & often said Lucy would ruin Samuel & his will gave a life interest only to F. & a nominal £40,000 more but all locked up in gun stocks. They supplied

[6] Priestley had already left before Samuel Jr. attended Warrington.

French Govt. with Firearms before '93.

F. gave all his papers (he never destroyed a paper but docketed all) to AB with desire to destroy all unnecessary—love letters. AB read some to M & she said "I had no idea you had been so much in love" she did not like it. 2 drawers full of letters about Mrs S. GF treated F very unkindly wanted every penny to spare of F income that he [needed] invest & make much of in the funds—war prices. GF in course of his life gave F £50,000 including marriage settlements, nothing by will. F had thought of building at [Warley] & living in a simple way. GF objected (M would never have liked living in a small way her ideas were not economical.) F was thus in much want of ready money.

[...] Barclay Mr. H. Gurney grandfather always begging for money for poor Barclay relations GF said he was not a golden calf. F built the large front piece of Duddeston. It was originally only a fishing house. Uncle John added to it. GF added to it, & F added largely to those addings. Mrs Lloyd (a Miss Farmer) lived there when a child.

Aunt Molly lived in GF town house (the Bank).

Biograph: chart of GF (John died 1799, born 1720, married 1746 wife's portion £1,600, admitted to partnership by Farmer, who previously had employed him) inherits 1754 from his mother an estate at Taunton.

F born 1753, a sister Mary born 1754 died insane.

Charles Lloyd of [Binley] married Mary Farmer, half sister became mad. [Hanbury] of [Egglestone]. David Barclay married for 2d wife another sister of Charles Lloyd that introduced the families.

F says Robert Barclay Mrs Hubert's father at ½ yearly Quaker meeting in London who introduced him to his future wife.

GF connected with Liverpool, selling guns for African slave trade. The gun trade died a natural death at the peace.

Mrs Schim' remarks about the gun stocks GF made in 1811. Mrs S so discontented she would not live at home. She wanted a separate portion of house entirely to herself for her own home visits & receptions F refused this. So Mrs S. would go from home.

It was Sampson Lloyd who when his son was sad & miserable said "tell thy friend Old Samp of all thy troubles & scrapes & I promise thee thy father shall never hear a word of it". He was the [farmer] who when the Butler taking [the big joint just] fell down & professed to be seriously hurt said "If thou wilt not sham I will forgive thee".

f.11: Aunt Brewin, Sophie Galton

A good Quaker is the perfection of Xtianity but it is in that a nothing

Keir was an unbeliever altogether. Watt had a respect for religion that was all. Darwin had reputation of being an unbeliever, he stuttered so that AB could not understand a word he said. (She was 11 years old.) He drank water enormously at night, had a Ewer always which he drank dry—was a teetotaller.

She was the most fascinating person—talent for caricature, exaggeration, consequent untruthfulness, wormed everything out of people & made her own use of it. $3\frac{1}{2}$ years older than AB. She would go to Mr Barbauld who knew the family & who took in young ladies to finish them. Then she got acquainted with the Earlham Gurneys & became so popular with them that she introduced the families more closely. Next to Mrs Beaver's establishment for single young ladies to introduce them to society, a fashionable establishment & remained there 2 winters.

All brothers & sisters had such a dread of Marianne that it was like a thundercloud the news each time of her returning. She always set them together by the ears. She often put herself into a pretended fit. (The doctor Vaux ridiculed them.) Mother never spoke about Marianne to any of the children till after F. illness.

Then she was 2 winters at Bath with Moravian family. Also staid with Mr Fowell Buxton's mother & took to riding. F. disliked it but she would ride & mounted a coach horse with blinkers & rode 30 miles a day with the coachman on another horse after her. Coachman was done! In this sort of way year after year dragged on—then she came of age under her GF will & F said she must go.

AB thinks Mr Sch must have been insane.

Old Watt's children Gregory & Jesse. Jesse went to school & in holidays to dancing school with AB. Sch trumped up the violent story about poor Jesse Watt. She (12 years old) wrote a little book with pictures describing that Jesse wanted to elope with a young man from the school & described all about the costumes &c. she was to adopt. She was a beautiful girl—

face swollen with sobbing. She went into a decline ascribed by M to this.

f.12.r: Aunt Brewin, Sophie Galton

Long-castre a French spy did the crayon of G. Father.[7]

(I & Mrs Fry 1825 comfits.)

Swiss governess – Mrs Sch GM (Allardice?) ~~who was a Cameron~~ misbehaved, she was married when 16 & ill and intrigued with a valet & employed her daughter Cameron to convey messages unconsciously between [...] poor Cameron had no protector & was assigned to Duddeston by GF F did not like to receive anything incorrect but did not. Cameron was overscrupulous horrified at what had taken place. At Duddeston was watched strictly. It was a time when "tokens" [outlied] of pence were in use. The servant got them for Mrs. S. who asked Cameron often to ask the footman for them. Cameron object. Mrs Sch: told M that Cameron was intimate with the footman & always stopped to speak to him. She was sent away from Duddeston. Her (C) father mistrusted Mrs Sch & believed in Cameron.

Other quarrels are described & docketed in a drawer. Miss Hant. Charles Lloyd. Permberton. Thornton Mr H. A Gurney nearly turned out from Earlham. Father & Mother. Hubert has stories about his own relation. Mrs Booth & AB. My father's [intrigue] with my mother. Miss Paterson—a long letter to M of 13 pages abusing her.

She set to work insidiously—began by praising then wormed round & repeated [proferred] remarks.

Duddeston never gossip — no small talk. AB never was a child none of them were. She was very unhappy—whatever she said was so caught up by Marianne or somebody that she never dared to open her [mouth]. In after life being of an active disposition she found a place in the home to occupy. Mrs Booth not. They were as silent as nuns. Mrs. B. spirit was naturally very high but became quite broken.

---

7 The known Longastre portrait is of Samuel Galton jr., not Samuel Galton sr. Perhaps he made two. He also painted Lucy.

f.13.r: Aunt Brewin, Sophie Galton

Grandfather began "compound" (F. He called it birthday money. Tertius named it compound, afterwards) £1 a year. Every year till they were 24 years old, F to pay income & let the sum accumulate.

T[heodore] was so noble spirited took after his mother more than any of them. He was much attached to [Emma] Darwin[8]—La Rochefoucauld says when a person has once been in love he [is] fully in love with love itself & often has another attachment.[9] Theodore's 2d[10] was unfortunate & [he] probably went abroad to get out of it. Her miniature was left in his desk after death, some thing which AB thinks a modest young woman would not like a man to have. AB returned them saying she[11] would probably prefer to have them returned to remaining ignorant of what became of them!

f.13.v: Aunt Brewin, Sophie Galton

Mrs Sch. delight to take away her own Father character, then Dr Darwin ordered M to take no wine but much animal servings food. Mutton slices at tea, mistake for muffin—exaggeration [...] of raw mutton. She said at a [...].

Father took Mrs Sch tower of London. Specimens of guns by different makers—Galton among rest. The showman pointed them out & said that rascal Galton he always furnishes us with beech instead of walnut. Dr. Smith said Miss S you may make yourself easy as to your father's reputation as no person conversant with wood could possibly have been deceived.

Tertius enquired previously[12] a beau.

9 false oaths before such a deception cd have passed. Theodore cut her out of will—left his money in 5ths. All thought no more of the matter, believed he would have survived her. The very day M[ary Anne] & Sch[immelpenninck] arrived first time after marriage news came of T[heodore's] death. T[heodore] had left keys with AB asking nothing to be opened except in certain particular cases. Tertius said "Don't open—Marianne

---

8 Emma Georgina Elizabeth Darwin (1784–1818). Unclear, but this seems most likely. She was Theodore's age.

9 'In their first passion, women are in love with their lovers. In all others they are in love with love', La Rochefoucauld, *Maxims*.

10 Mary Gribbins.

11 Mary Gribbins.

12 Possibly 'seriously'.

is not named in the will"—he had seen Theodore's will & begged him not to do as said—making will is a solemn act & it is a pity to let any feeling be perpetuated by such an act. AB disliked & said not doing this looked as if we were guilty. They referred to F[ather] who said "pray don't ask me Theodore has spoken to me about it". Theodore said she never had been a sister to me. The Schims went then will was opened, Tertius was frightened & begged AB to communicate with Mr. Schim which she did. AB wrote a nice letter to him to say Mrs S name was not named in the will. Mr Schim wrote a most civil & cordial reply—this was moonshine. She then reiterated all her old demands about settlement matters & the row came. Mr Sch had kept a journal writing everything in familiar conversation to him—& drew up a statement.

There had been great troubles about Mrs S marriage settlement as to what F was to assign her (Mrs S always represented her F[ather] had cheated her, most unjustly). Mrs S was in the habit of sending for her mother to come to her room, just like a servant—& she did!

f.13.v

Draft of Memories section about Galtons.

f.14

Several very curious qualities, all more or less hereditary have been found in the families of the persons under review. I will not dwell upon the better known tendencies to music drawing mechanics and the like but will specify one that strikes me as very curious. It is the love of figures & tabulation.

[Guy[13]] In one instance the present possessor has it in a remarkable degree as abundantly testified by his works & he states that his grandfather but not his father had it also. This family happened to consist of but a few individuals [...] does not afford opportunities for much further evidence of heredity.

[Galton] But I know of another one family which also falling within my field of inquiry in which the same love of figures & tabulation has been hereditary, in a most curious manner. The

[13] Probably the statistician Dr. William Augustus Guy (1810–1885).

f.15

grandfather, tabulated every transaction in his household. I have had the opportunity of seeing bundles of his papers of that character. As an example, one gave a tabular view of all his successive servants in different capacities. It was a long sheet of paper ruled neatly in columns with red ink one column to each class of servant, and the whole looked like those large synoptic sheets of contemporary history, in long parallel columns. He had a large family in some of whom the same instinct was curiously transferred. One of his children was a lady who seemed possessed with almost a monomania for tabulating. She had peculiarly large fortune or duties but the number of ledgers & ruled books in which she indulged, almost surpasses belief.

Her ordinary inkstand was shaped like a trefoil, holding black, red & blue inks & she used all these in ordinary correspondence to say nothing of her tabulated entries. Her jugs and the like were all marked with her initials, purpose for which they were to be used, & the name of the room in which they were to be kept. And none of this elaborate machinery was of real use on the contrary it hampered. But it was the creation of an imperious instinct which required to be satisfied. I could mention others of that generation but confine myself to one of her brothers, the father of the person within my lists.[14] He loved figures & tabulation. Amongst other things sliding rules abounded in his house which he used continually for quaint purposes. These at one time when his sons & daughters were young, he & his wife calculated the number of bottles of sherry & champagne, of fowls, tongues and all the rest were always calculated by his sliding rule. He had appropriate card faces on which lines were drawn corresponding to the various eatables & drinkables & also a zero line. Then he set the number of expected people, in the rod of the sliding rule to the zero line & read off the necessary number of bottles of sherry & the rest at the several lines. He also published certain tabular charts of considerable elaboration & in advance of what was customary in his day.

His children very generally possessed that same love to tabulation. ~~and in some extent of figures~~ One daughter had it in combination with a passion for perspective drawing, wholly innate, another daughter if not two have also shewn the love

[14] Tertius Galton.

of tabulation in a marked degree among other things being especially fond of genealogical tabulation. One son is singularly methodical entirely individual in his ways and keeping everything ruled & docketed in a most curiously perfect order, though utterly without business education, another son has the same instinct to a considerable degree. And the contributor of this information states himself to be possessed to an inconvenient ~~disagreeable~~

f.19

extent by this same curious natural desire.[15] It is stronger in him than subserves any ~~useful~~ end that he happens to want and it encourages a form of mental misapprehension, namely one that is more narrow, hard and precise than is true to nature.

I have entered at some length upon this example to shew what curious mental qualities are to be found and how hereditary they may be but I am bound to say speaking generally it does not seem to me that only in particular aspects is so often hereditary as the more general tendency to attach oneself warmly to some object or other. Then though insanity is strongly hereditary monomania with respect to a special subject is not so.

f.20

In endeavouring to dissect my own nature, and to account for it I am astonished at its mongrel antecedents, for I have many relatives of mark but all in different ways, whose several peculiarities have been curiously mixed up among their descendants. Hence notwithstanding the characters of the latter are diverse, all the elements of which they severally consist admit of being traced. If it were proper to write fully upon these I am sure that a history of the Galtons would be found unusually instructive from an hereditary point of view. ~~I confine myself to some of the incongruous elements of which it is composed.~~ My Grandfather Samuel Galton FRS and an early member of the Linnaean Society (memoir on colour in the Phil Trans & a small book on birds) was a wealthy Quaker ~~much mixed up with the~~ of a disposition full of contrasts and

[15] Francis Galton.

closely connected with the literary and scientific society of the Midland counties towards the end of the last century, which had free thinking, philanthropic & industrial tendencies (see description of him in the memoirs of his daughter, Mary Anne Schimmelpenninck I.250, small edition 1858). His wife had force of character & intelligence & a peculiar stateliness of manner combined with a rather pedantic turn of mind and many singularities. She was descended through a very mixed line maternally from Barclay the "Apologist" and was sister of Captn Barclay of Urie, a man of adventures, one of the strongest in England, the pedestrian of 1,000 miles in 1,000 hours, noted in coaching & hunting annals & in later years as an agricultural reformer. His memoirs are published.

For example every body boxed in those days, and Jackson whom Barclay greatly patronised, was the champion pugilist. They continually boxed together on not unequal terms, Barclay being superior in sheer strength but Jackson in skill. I will relate another instance of his power. When he was past 70 he was dining at my father's house, and being asked about

f.21

his strength replied by desiring my brother, who weighed [ ] stone, to put one foot in his hand, which he laid on the ground, and to steady himself by resting the tip of the finger on his shoulder; then Barclay quietly, still sitting in his chair, lifted my brother on to the dinner table with that one hand. But he strained his arm in doing so, being old & out of training (his Memoirs are published). His father & grandfather were both accredited with almost fabulous muscularity.

(Theodore. One who had the instinct of a traveller & died of plague in early life in the Lazaret at Malta & another who with great love of locomotion & remarkable organising power & endurance of fatigue.)

The character of two sons may be alluded to & one daughter mentioned by name. Of the sons one was reputed to have high gifts as well as the instincts & aptitudes of a traveller. He died young of plague in the Lazzaret of Malta, Another was characterised by love of society, energetic organizing aptitudes, a love of rapid travel & powers of enduring fatigue.

Of the 4 sons & 3 daughters of Samuel Galton & his wife, I shall only speak of one, though others might deserve mention, namely a married daughter, Mrs. Schimmelpenninck, authoress of Memoirs of Port Royal, Principles of Beauty &c. She had great sensitiveness & religious fervour, joined to a strong intellect. I have lately heard no less than three very notable people who knew her in her mature and declining years speak of her in expressions almost of adulation. She is said to have then exerted extraordinary influence for good over her acquaintance. ~~But I speak from hearsay, having had no personal knowledge of her myself.~~ She was almost wholly separated from the rest of her family after she grew up. The subject of her separation is spoken of in those memoirs and [fanciful] though it be, I think it right to use this occasion to say something about it from an entirely different point of view to that of the author of them. It is just possible that the traditional sense of harm long since done, may linger in other families whom this book may reach & be thereby effaced. It appears to me that in her early womanhood, Mrs Schimmelpenninck must have passed through a morbid psychological phase not uncommon to medical experience, in

f.22.r

which the moral sense is temporarily perverted. Grounds of this belief may be gathered from her autobiography in Vol. I of the above mentioned work which extends to at 15.[16] The Vol. II is written by a friend of later years and includes the remainder of her long life, but the critical period of which I speak, has had no adequate historian. She herself describes (I.267) how the religious faith of her childhood was wrecked in her girlhood, how her affections ceased (309) and how her character changed for the bad (308). In page 268 we may trace the commencement of the phase to which I allude, namely the development of a spirit of subtle persistent and ruthless mischief making, the more dangerous because her charming manners won confidence and she had a large and scattered acquaintance. This extended over many years preceding her marriage at 28 which was childless & was the real cause of the complete separation of the family from her [...] domestic agonies endured for many years, but which I may only glance at here, she was extruded (as far as possible!) from her home, long previous to her marriage at 28 (which was childless). Neither was subsequent reconciliation feasible, because misrepresenta-

[16] That is, age 15.

tions remained disavowed & the slurs abided. Her biographer, taking her part, in II.74 seems wholly ignorant of this ~~very important part of the story~~.

The next generation 7 sons & 7 daughters is that in which I come to [include] my cousin Capt Douglas Galton FRS distinguished for official activity in his various capacities especially in connection with sanitary improvement of hospitals [...] once secy of the Railways Department, then of the financial Departments of the War Office & now Director of Public Works.

~~More than this I may I cannot without speaking of persons whose life is not strictly public~~ On the maternal side of the Strutt family noted for the mechanical aptitudes of its founders & of 3 of 4 other members among his numerous descendants.

There is much extreme hereditary interest to be traced in this generation but I could not speak of it without violating the [confidences of] private life

Chapter 2

Galton's Report on Damaraland

The text of this report has been reconstructed from draft copies in the diaries of Galton and Andersson, neither of which contains the complete text on its own. The original was once held in the Cape Archives, with a copy placed in the Namibian Archives, but both institutions can no longer find their copies.

Many years ago the Damaras lived about the Kuisip, at Glenelg Bath, & at points even more to the south than [that], while J. Africaner was a chief of secondary importance near the Fish River. He fought with, & conquered neighbouring tribes, & gradually increased in power. At this period a distinct race of Namaquas, who lived on the Kubakop & who were called the red people, asked his assistance against the Damaras.

He gave it, & conquered them. As he became stronger he made war against them on his own account, drove the Damaras before him, & established himself at Glenelg Bath and at Cibris[1] where he was when Sir J. Alexander saw him. Next, about the year 1839-40 he pushed on to Concordiaville, where certain missionaries settled with him. Here Mr. Hahn, Mr. Kleinschmidt and Mr. Bam were witness to, and assisted in, the gradual establishment of peace between him & the Damaras, a peace that was kept most satisfactorily by all parties during 1843, 1844, 1845. Trade between them flourished greatly, and Concordiaville had blacksmith shops where iron things were continually being forged for the Damaras, who brought there their cattle in exchange & the place had the appearance of a crowded, bustling market, the whole year round.

[1] Tsebris, west of Rehoboth.

Early in 1845, certain East Damaras (Ovampantieru), under [Kaitchekenge] came to Jonker asking his help against some West Damaras (Ovaherero) under [Katusieko], for one of their tribes had been utterly destroyed by him, having been decoyed to his werft, under pretext of a feast, & there murdered. This assistance Jonker at first refused, but being at the same time urged by pressure of debt (800 cattle to J. Morris) he afterwards gave it, & started with them. He could not find [Katusieko] but passing in his way by a werft (under [Kashena]) and who were utterly unconnected with the dispute in question, he nevertheless fell upon them, and, though they fled, he followed them, & shot them down men & women including [Kashena], and took prisoners of the women & children, and seized their cattle; returning, he fell upon two other werfts in precisely the same way, seizing their cattle, but their inhabitants—it being night—escaped. This occurred about 5 days journey NNE from Concordiaville. This was told to Mr. Hahn, in the first instance, by April, Jonker's servant, who was thoroughly cross-questioned; it was afterwards confirmed by many witnesses, and is a matter of public notoriety. On that, certain of the red people, under Cornelius, having heard that Jonker had broken the peace, determined to break it too, and fell upon some of [Kaitchekenge's] people, under his brothers [...]. They came in a friendly way, as guests, and as they were giving them milk, they commenced the attack shot [...][2] and the flying[3] woman & children were massacred by Berg Damaras belonging to Jonker. This latter proceeding, whether honestly or not, Jonker expressed his regret at; but he returned to Concordiaville with at least 1500 robbed cattle, out of which he paid his debts, which amounted to 800 on his own part, & probably much also from the rest of his tribe (this number 800 either J Morris himself or else his partner Mr. Dixon[4] informed the late Mr. Scheppmann, who communicated it to the missionaries here; it was besides [a] matter of notoriety.) Jonker immediately pursued that system which he had begun with such success. In September 1846, he went with most of his men to the NNW. He was received in a very friendly way by a tribe under [Oandjizuva]; he asked their hospitality & received it, but on a pretext, known to have been utterly un-

2 Text has been omitted here in both Andersson's copy and the original in Galton's journal.

3 Sic.

4 Dixon, Peter Daniel, see page 356.

founded, of some of his cattle having been stolen, he rose in the night, slew the chief & many others, & stole the cattle. He then went on & destroyed two other werfts in succession, killing what men, women & children he could. A second time he returned home enriched with plundered cattle; since that time this system of outrage has continually gone on with literally no intermission, except when the season rendered such expeditions impracticable. Mr. Kolbe informs me, that, about a year since, when in Cape Town, he had an interview with his Excellency the Governor, & that he (Mr. Kolbe) abstained from bringing the facts before his Excellency's notice because the scene of plunder had not at that time been near his station, & he dreaded being brought into collision with Jonker, which he certainly would have been had Jonker suspected him of giving information to Government in any way tending to his detriment. The expeditions of Jonker since the one last mentioned have been exactly similar in character to it; they have uniformly been accompanied with the greatest barbarities: the arms and legs of women and children are cut off when they are alive in order that the iron rings they wear as ornaments may more easily be got off. Children are slaughtered & mutilated for a savage thirst for blood. And there is a peculiar mixture of cowardice & ferocity in all their onslaughts which is thoroughly disgusting & abhorrent. I therefore pass over the time intervening between the last mentioned & one that is the immediate cause of a barbarous & audacious outrage upon Mr. Kolbe's station by Jonker which occurred at the time of my arrival in this land & of which I can from knowledge obtained on the spot give the particulars with all confidence.

In April 1848 Jonker was on the road towards Walfisch Bay and at that time many Damaras, including Kahitjene, were lying in that direction. Jonker came to him & Kahitjene said (according to Jonker's statement to Mr. ——) that he had himself no cattle to spare him for food, but that there were tribes, some of Jonker's immediate enemies, a little further on with many herds. These he immediately attacked, although it was proved that they in fact had never had anything at all to do with him. He killed four chiefs, and at least 40 people, & stole their cattle. In the medley[5] some of Kahitjene's own people were also attacked. For this Jonker expressed his regret, but Kahitjene was not to be convinced and attacked

[5] sic

Jonker that same night with but little success, while Kahitjene's people suffered severely. Since then Jonker seems to have determined to utterly exterminate Kahitjene's tribe.

Now this year a few of Jonker's cattle had been stolen by the [Ovampantieru] (or East Damaras) & in retaliation instead of following the thieves, Jonker's people under Jan Jonker together with [Katchikenga] & his people, attacked an innocent tribe of their Damara allies who lived close by Concordiaville, killing the chief & others. Mr. Hahn passed the scene of attack a week after it took place and heard the history on the spot. The fugitives went to Kahitjene who in his turn attacked [Katchikenga]. On that pretext Jonker has 3 months since fallen upon Kahitjene while he was living close by Mr. Kolbe's station. Having scattered them he then attacked the village of Damaras that he had collected round Mr. Kolbe's house who were entirely dependant on him & who were utterly unable from want of arms either to give or to resist offence.[6]

Mr. Kolbe's house was not actually besieged though bullets were flying about in all directions; this lasted all night. In the morning Jonker came in a drunken state to Mr. Kolbe's; he was unable to give any consistent account of his conduct further than implying that he would do what he liked & would not be interfered with. The place is now quite deserted & Mr. Kolbe is living here at Mr. Hahn's station about 9 miles[7] off.

On his return Jonker fell upon [Katchikenga], the very person in whose cause he professed to fight, & utterly destroyed the tribe except the chief & a few women who escaped.

Thence going home Jonker fell upon the Damaras who were living round about his place & under his own protection, and served them exactly in the same way & took their cattle.

The audacity of Jonker's outrage upon Mr. Kolbe's station becomes much more striking when the peculiar customs of the country are considered. The natives have not the slightest idea of, or care for, territorial right. The country is very wide for the population & any man of any nation or colour who happens to encamp or settle by a pasturage is considered, so long as he remains there, the owner, and another tribe passing

6 Additional erased text in Galton's journal: ~~Two children of these were ripped up under Mr. Kolbe's eyes, a third had its eyes stabbed out, another its hand cut off.~~

7 Andersson: I think instead of 9 miles to must be twice that distance; for it took us on coming here with the waggons 7½ hours.

that way always asks permission before they venture to feed their cattle there. Now that the missionaries have built houses many Damaras whose tribes are broken up or who are starving for want come & settle by them, they get a little work & they consider the missionaries as their chief whether they like that honour or not. When he lays out gardens, the gardens are respected by them & they never think of encroaching upon them; and when they themselves want garden land they never think of taking it without it having been formally given to them by the missionary. Thus these Damaras become in a way the retainers of the missionaries, and I must give my most willing testimony to the excellent manner in which they keep order amongst them. It was the retainers of Mr. Kolbe whose huts Jonker attacked, robbed & destroyed. These proceedings have thrown the whole country into a most lawless state. Any idea of us now making a direct route to the lakes lying, as it would, right across the lands ravaged by the Namaquas, is quite out of the question. Both natives and Namaquas would resist such an expedition. The Damaras believe that all white people are the immediate allies of the Namaquas. They know that their cattle are stolen to settle the white traders, and thus they have much reason to think so. I have done all I could to get an interview with the Damara chiefs, but without success, as again another attack has occurred which has quite scattered them & made them desperate & thrown serious difficulties in my progress.

As soon as I had satisfied myself fully that there was no exaggeration in these accounts of Jonker's proceedings, I felt it my duty to act here upon the instructions you furnished me with as being so entirely within the scope of their intent, and indeed verbally so, for Jonker himself can best be described as an emigrant Boer[8] & this country, as that in the neighbourhood of the lake, for the Ovampantieru (whom I have already mentioned) & the [Ovationa] are the only intervening tribes. I was assured, also, that Jonker felt great alarm at the displeasure of the British Government, and was the less careful here as he supposed this to be a country that England would never trouble herself about, and therefore I wrote a letter to him—and at the same time I sent a short friendly message to the four principal chiefs asking them if they would come here & meet me & if so I would appoint a day for them to come.

8 A definitely false claim, and one that Galton is unlikely to have made sincerely.

I had hoped that Jonker would have avowed his intention of making, if not amends, at least peace with them & that that welcome news would have much inclined the Damaras to the British. However, immediately on Jonker receiving it, an expedition of his men combined with Amiral's people & the red people[9] were on the point of setting out. He sent his answer a round of 300 miles, most probably guessing that I should wait for it before seeing the chiefs; and the expedition set out. It is by far the most extensive one that has yet gone. Eleven of Kahitjene's remaining fifteen werfts were destroyed; ten others belonging to [Katjikura] & 5 to a third chief. This would represent about eighteen thousand head of cattle. On my receipt of Jonker's answer, of which I enclose an abstract, (the original being very long couched in very humble terms and not very intelligible I did not think it worth while to send a copy) I wrote again to him the second letter. He was then some distance away dividing the spoils & I have as yet received no answer. The cattle that the Namaquas steal is for the most part sold to traders who come from the colony into their country. Very much of it goes to buy horses. The horse distemper is very severe here as I know from my own experience, for the remaining horse that I had myself, & even 3 mules, out of eight, died during the first fortnight I was here & another mule was on the point of death. I therefore quite credit what I have been assured that for any horse that lives, at first coming into the land, & none hardly lives more than one or two years, nine die. Now the current price of a horse in the part of Namaqualand where they are bought is from six to ten oxen, therefore each efficient horse represents about 80 oxen. Now Jonker alone has at his command quite fifty horse & they therefore must represent a constantly renewed expenditure of 4000 oxen. This gives an idea of the fearful extent of these robberies. The cattle must all be stolen as no trade is now carried on between the Namaquas & Damaras. The horses, I should add, are necessary to cattle robbing for as soon as the firing commences the oxen run wildly straight away & no man on foot could possibly overtake and turn them. There has never been shown on the part of the traders the slightest scruple about taking stolen cattle in payment.

⟨I have written this much in the case that the British Government should think it either right or expedient to interfere in any way in this distant land. Further information as to the

[9] Nama under Cornelius.

number of cattle stolen could certainly be obtained at Nisbett Bath (sometimes put down as Warm bath on the maps Lat. 28 20 Long 18 30) through or close by which nearly all the droves both from the Red people & Jonker's pass (it is also as I hear, though I cannot speak confidently, the only practicable road for them). Whether that information would be honestly given by any person native or stranger now on the spot is I understand very doubtful. It would be also important to know if any of the great number of slaves taken by the Namaquas are exported to the south of that place. If they were and if any of them had learnt to speak Dutch they would be invaluable interpreters to any persons who had to deal with the Damaras. Enquiries might also be made at [Pella] and at [Kock Toulein] (Steinkopf) through or near which many of the droves pass. If you would kindly permit me to do so I will take the liberty of making a few suggestions about a course which I conceive that Government might adopt if it thought proper to interfere at all in the case. I am quite convinced that the Namaqua obtain their superiority over the Damaras, who are in every respect the finer race, by guns & by horses only, and as those are obtained almost entirely by exchange for the stolen cattle, and again as this trade admits of a very easy check at the places I have mentioned especially if Bondel Swartz tribe could be induced honestly to cooperate, I conclude that much effect upon the Namaqua would be produced if that check was for a time resorted to. The Namaquas would then feel their entire dependence upon the country to the south for supplies & the exhibited power of the Cape Government in stopping these would soon I should conceive reduce the Namaquas to some obedience. From what I hear on all sides and from what little I have seen, they are a most inferior race and not at all likely to shew themselves careless of opposition & acquire a national spirit by experiencing it. Those who are acquainted both with the emigrant Boers & with them assure me that they are not to be compared. I wish that I could recommend diverting the cattle trade to Walfisch Bay. It certainly borders on the Damara country, though itself claimed by Jonker, it has great recommendations for a large salting establishment. Fish abound in the Bay, but on the other hand the confidence of the Damaras is so entirely lost that a cattle trader even if he had good interpreters could not for a long time succeed.

(Mr. Haddy the missionary at Jonker's place will probably be at Cape Town some time before the receipt of this letter, to him also I beg to refer the Government for information)

I enclose a map in which I have dotted the parts that I have not visited, it gives however a better idea of the country than the printed maps. A small map made by Mr. Hahn & which came into the possession of Government is I find in no way accordant as to its principal features with Mr. Hahn's information about the interior of the country. He was anxious to inform the missionaries of the position of the different stations here, & he therefore traced off the outline of McQueen's proverbially incorrect & theoretical map, but the only one that he happened to have by him and on it he dotted the various stations. It has acquired a very undue importance as giving the geography of the interior. Mr. Hahn has himself ridden to one of the places marked on the map as particularly mountainous & found it quite flat.⟩[10]

If the Damaras were possessed of fire arms, & if the Namaquas had no horses, there is little doubt that the former would defend themselves; if, however, they had entire superiority over them they would certainly retaliate all the cruelty they had received and would abuse their power to the full as much as the Namaquas do now. There is no union among the Damaras; they never will act in [concert] or under command; this the Namaquas do, for they have some notion of discipline.

⟨I have urged Mr. Hahn upon whose pains-taking enquiries & great information much reliance could always be placed to write by each post to Government giving an account of what passes here. I conceived that such information would be very valuable at the Cape, and if the missionaries here do not judge it inexpedient that one of their body should thus regularly communicate the affairs of this land he will do so. There is a post established here by missionary cooperation, every four months, it has not been very regular from mismanagement at ——— where the messengers toward & from the Colony exchange letters. I mention this arrangement as cases might occur in which it would be of convenience to Government. If Jonker does not intercept all communication exertions will be made to reduce it to better order. The letters take about 3 months on the road & pass by Mr. Morgenroed a merchant. The agent of the Rhenish Society at Cape Town will know the times that it leaves Cape Town.⟩[11]

Extracts of Mr. Haddy's letter:

---

[10] Additional text from Galton's folio diary, GALTON 1/3/2/1.

[11] Additional part in Galton's folio diary, p. 34.

> "What amount of blame may be attributed to the Damaras I will leave others to judge who may know them better, but it is too clear that a reckless spirit of robbery & murder has come upon the Oerlams & Namaquas generally along the entire boundary line. In some instances, it appears that they do not even feign a pretext to give a degree of plausibility to their inhuman practices. 'The Damaras have cattle & we have the power to rob them' seems to be in many cases the sole assignable cause. This is, however, no new thing in the earth, &c.&c."
> July 25th 1849
> To Mr. Hahn.

Again, from another letter:-

> "It is in the highest degree painful to think that the Damaras in general are so dreadfully slaughtered by the Namaquas & Bastards of this country, & at the same time they should be so much disunited & so decidedly opposed to each other, so that there is no hope of their combining their forces for a common defence. It is moreover evident that the plundering expeditions of the Hottentot races against the Damaras have returned after being checked for a few years with all their original or with increased fury so that it is becoming quite a passion & a business from the Orange river to this latitude to prepare for, aid in, or in one way or another promote the spoliation & murder or enslavement of the Damaras. I have long feared that the chief of the Bondelschwartz is not indifferent to these iniquitous proceedings but I have no facts on which to found a direct and specific accusation. And the question is not who is guilty? but who is guiltless?
>
> A few that fear God &c.&c abstain ... but my patience is exhausted when I think of so fine a class of men capable of learning any thing, of having as fine a polish mentally & physically as the inhabitants of Paris & Berlin, falling before the little minded, obstinate-headed, lazy, piggish Namaquas."
> 17 Oct 1849 to Mr. Hahn.

Similar sentiments are expressed & implied in a letter dated 12 Jan 1850.

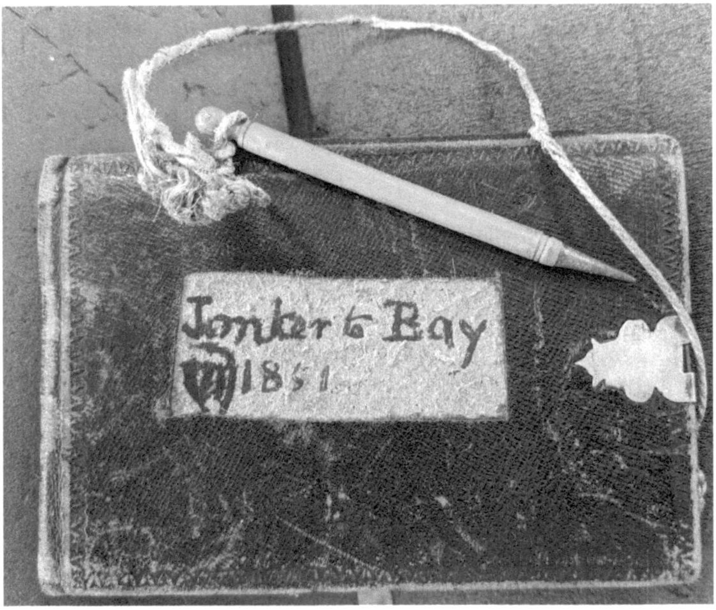

Figure 2.1: Galton's Sturdy South-West African Notebook VI. The pages were 'metalled' with bone-dust and paste.

Figure 2.2: Hykomkap on the Swakop River, by Thomas Baines, 1861.

Chapter 3

Wreck of the *Dalhousie*

Extraordinary Wreck of The "Dalhousie."[1]—The fine and well-known ship the *Dalhousie* of London, Captain Butterworth, commander, suddenly foundered in the Channel, with terrible loss of life, and the destruction of a very valuable cargo. The circumstances of this wreck are of a most extraordinary character. The *Dalhousie* was a fine Indian teak-built ship, of nearly 800 tons. She was launched at Moulmein in 1848, and was therefore a first-class vessel, in good and sound condition. She was the property of Mr. Allan, of Leadenhall Street, and was chartered to Messrs. Fry and Davison, of Fenchurch Street, as one of the "White Horse line of Australian passenger ships," but in consequence of the great demand for tonnage to those colonies, upon this occasion she was chiefly taken up for freight. Her cargo comprised every description of merchandise, estimated at upwards of 100,000*l.* value. Her master, Captain Butterworth, is said to have been a most experienced navigator, and to have passed many years in the owner's service.

She cleared from the docks at Blackwall on the 12th,[2] and proceeded down the river to Gravesend. Her passengers were—Mrs. Underwood and three children,[3] Mrs. Simpson and three children, and Miss Radford. Mrs. Butterworth being anxious to accompany her husband, Captain Butterworth,

[1] *Annual Register 1853*, Rivington, London, 1854, pp. 134-7.

[2] October 1853.

[3] Her husband Mr. Underwood was on board too, leading to an interesting legal case turning on whether the two had drowned simultaneously, or whether Mr. Underwood had for a while survived his wife and children, with important consequences given Mrs. Underwood's will. The court ruled that, absent definite evidence to the contrary, they died together.

to Plymouth, went on board with three of her sons for the trip down Channel; the eldest, however, on the ship reaching the Downs, went ashore at Deal. Altogether, the passengers and crew consisted of 63 or 64 persons. On the 18th, the ship's departure from the Downs was telegraphed to Lloyds, and the next that was heard of her was about 5 o'clock on the following afternoon, when a boat reached Dover from a brig lying off in the roadstead, reporting the appalling loss of the ship, and landing the only survivor, Joseph Reed, a seaman—every other soul on board having perished with the vessel. The protest of the sole survivor of this melancholy catastrophe states, that the ship arrived in the Downs on the 17th instant, and lay there detained by adverse winds and stormy weather until the 18th. During that time the ship rode easily through a fierce gale and a heavy sea. She left the Downs at 7 A.M. on the 18th, in a fresh breeze; after passing Dungeness, the wind increased and sail was taken in; when off Beachy Head, it blew a heavy gale. "At 2 A.M. I took the helm, from which position I had the means of observing everything which took place subsequently; and at 4 A.M., the fore and maintop-sail were double reefed, and the mizentop-sail stowed. The wind was then blowing a gale, accompanied by a heavy sea, in which the vessel laboured greatly. Up to that time she had shown herself, to the best of my judgment, a lively manageable ship; and during the gale in the Downs, she had behaved remarkably well. The pumps had been sounded night and morning, and she was quite tight, and upon trying them as usual at 8 P.M. (previous night), no water (excepting that always in the bottom) could be found. Towards 4 A.M. the ship began to lurch deeply in the sea, going a long way over on her broadside, and she seemed unable to recover herself. When she rolled I began to suspect that there must be a considerable quantity of water in her, and I mentioned to one of my shipmates that the vessel seemed like a water-logged ship. Shortly afterwards, the starboard quarter boat was carried away by a sea; and about 5 A.M. the crew, by the commander's orders, commenced throwing overboard water-casks, sheep-pens, and other lumber from the deck. While they were so occupied, the ship gave a violent lurch to starboard, and a heavy sea at the time breaking over her to leeward, washed overboard the longboat, which was full of live stock. The weather was then getting worse, and the ship was kept hauled to the wind on the port tack under the double-reefed fore and maintop-sail, foresail, and foretopmast-staysail, the foresail being hauled up and the maintop-sail lowered down on the cap. She continued to lurch violently, and at half-past 5 A.M. she rolled right over on her starboard beam-ends, and remained in that position with her mastheads in the water, lying at the mercy of the sea, which then made a clear breach over her, and washed away the larboard quarter boat. A great many of the crew took refuge in the maintop, and I got outside the ship on the weather quarter-gallery, it being impossible to stand on deck. A sea broke into the maintop sail, and carried away the maintop-mast with all its gear, and likewise washed four men out of the maintop. Captain Butterworth, the chief and second mates, the carpenter, cook, and some of the crew, joined me on the weather quarter; and they dragged through the gallery window, four passengers, consisting of a gentleman, his wife, and two children, who took refuge with them." This declarant and another seaman also succeeded in getting out of the water a young lady who had come out of one of the poop cabins. "I lashed her to a large spar, and placed her with the rest of the party on the gallery. Immediately afterwards, a large sea broke over the ship, which washed off the gentleman above-mentioned, with his wife and children (four in all), and they perished together. At this time a schooner was observed about half a mile to the eastward, bearing down upon the wreck. The vessel was at that time settling fast in the water, and it was evident that she could not remain afloat many minutes longer. I cut the lashings of the spar to which the young lady had been made fast, in order to give her a chance for her life." As the spar went adrift, Captain Butterworth, the second mate, and one or two of the seamen quitted the sinking ship, and held on to the spar, in the hope of saving themselves. Declarant being left on the quarter with the cook and carpenter, many of the people had by this time been drowned, but others remained holding on as they best could on the weather side of the wreck. She lay thus for about 10 minutes after Captain Butterworth had left her, and then sank, going down head first. "I scrambled from the quarter to the mizen-mast, which I ascended as the ship sank. I found the surgeon in the mizen-top, and we went up together in the mizen crosstrees, when we were submerged. I lost sight of the surgeon, and I swam to some deals which were floating about. I got hold of one of them, but shortly afterwards I saw near me one of the chocks of the longboat, capable of affording me better support than the deal, which I therefore left and placed myself on the chock. The schooner was then within shouting distance, being then about 100 yards to the leeward of me, and I hailed her, begging her crew to go about to windward, and afterwards drift down among the *Dalhousie's* people, of whom several were still alive, and might thus have been picked up and saved with a little exertion on the schooner's part. The only response which I could hear from her was given by some person on board, who told me to "swim to her," but she was drifting to leeward much faster than any man could swim, and she shortly afterwards stood away to the S.W., and left me and my companions struggling in the water. I watched her for nearly two hours afterwards, but she at length disappeared, without having (so far as I could see) made the least effort to save any of us, although the schooner was close to the *Dalhousie* when she foundered, and the crew must have seen her go down. In the course of the morning, several other vessels passed near me, both going up and down Channel without seeing us. My companions gradually perished one after the other, and I was repeatedly washed off my frail supporter." The poor mariner was afterwards rescued by the brig *Mitchel Grove*, of Littlehampton. It is supposed that this sudden calamity, which occasioned a great sensation at Lloyds, originated in the starting of a "butt," or that the vessel had struck upon a wreck or spar floating in the Channel. It is proper to add, that the alleged inhuman conduct of the schooner's crew is denied.

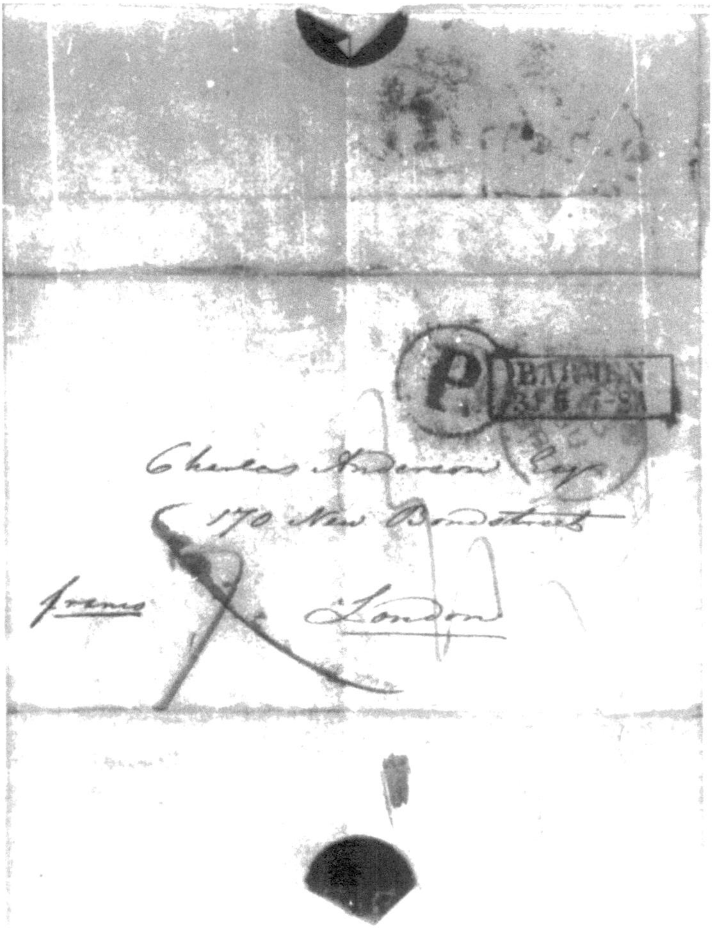

Figure 3.1: Charles Darwin to Francis Galton circa 1856, containing a query about the domestication of animals.

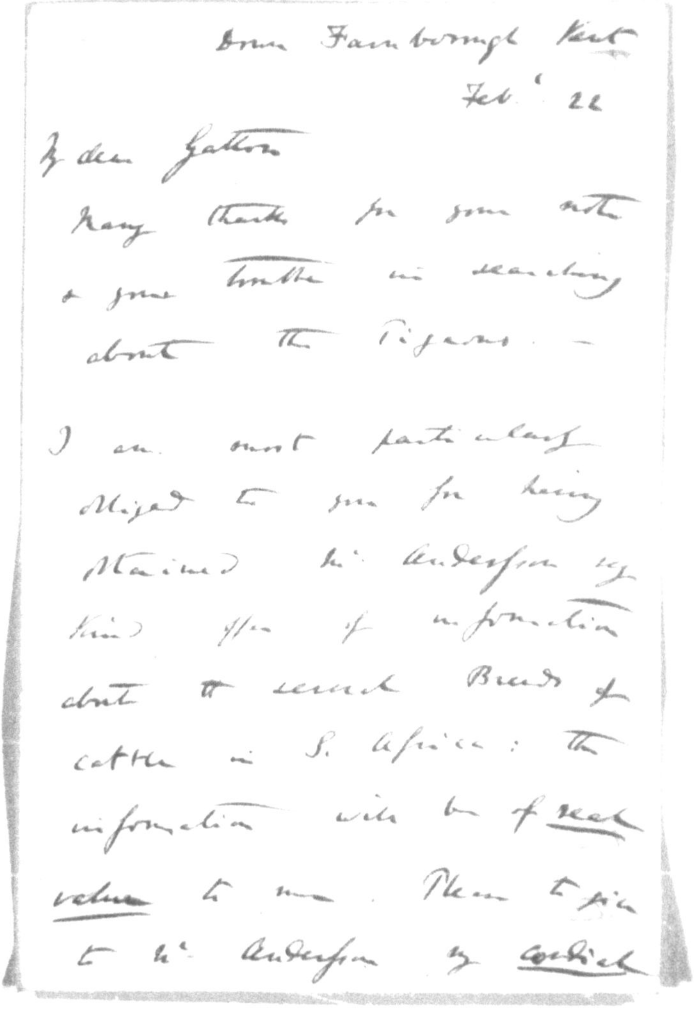

Down Farnborough Kent
Feb. 22

My dear Galton
Many thanks for your note & your trouble in searching about the Pigeons. —

I am most particularly obliged to you for having obtained Mr Anderson's very kind offer of information about the several Breeds of cattle in S. Africa: the information will be of real value to me. Please to give to Mr Anderson my cordial

Figure 3.2: Charles Darwin to Francis Galton: Page 1.

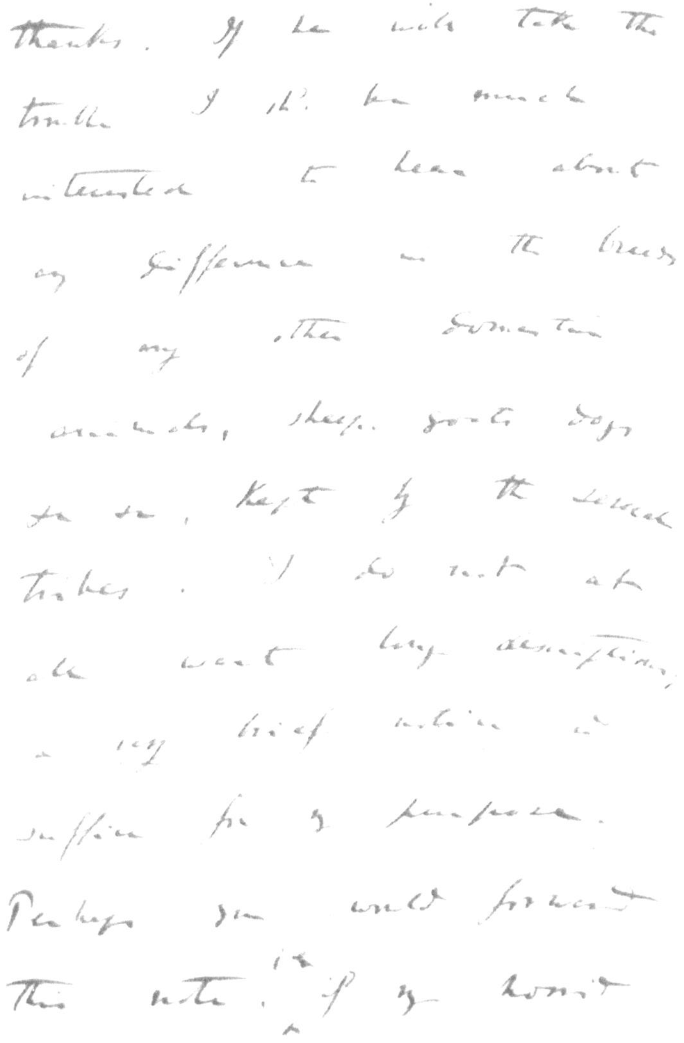

thanks. If he will take the trouble I sh. be much interested to hear about any difference in the breeds of any other domestic animals, sheep, goats, dogs &c &c, kept by the several tribes. I do not ask for any long descriptions, a very brief notice w^d suffice for my purpose. Perhaps you would forward this note, it if you think

Figure 3.3: Charles Darwin to Francis Galton: Page 2.

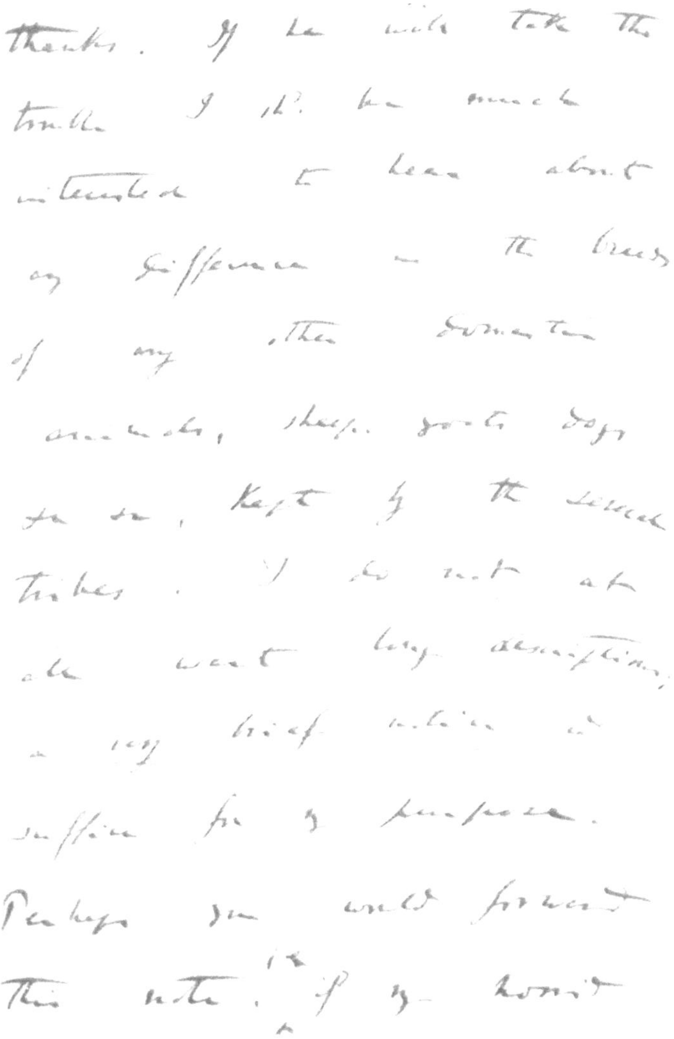
thanks. If he will take the
trouble I sh^d be much
interested to hear about
any difference in the breeds
of any other domestic
animals, sheep, goats, dogs
&c &c, kept by the several
tribes. I do not ask
nor want long descriptions,
a very brief notice w^d
suffice for my purpose.
Perhaps you would forward
this note, if you think it

Figure 3.4: Charles Darwin to Francis Galton: Page 3.

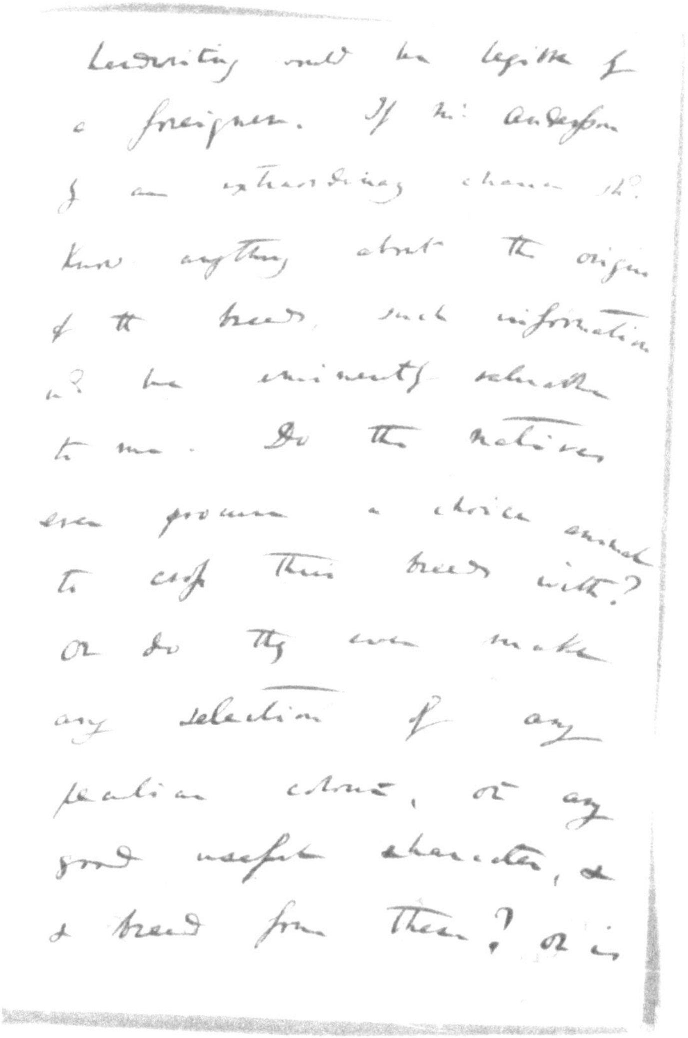
handwriting would be legible to
a foreigner. If Mr Anderson
by an extraordinary chance shd
know anything about the origin
of the breeds, such information
wd be eminently valuable
to me. Do the natives
ever procure a choice animal
to cross their breeds with?
Or do they ever make
any selection of any
peculiar colour, or any
good useful character, &
& breed from them? Or is

Figure 3.5: Charles Darwin to Francis Galton: Page 4.

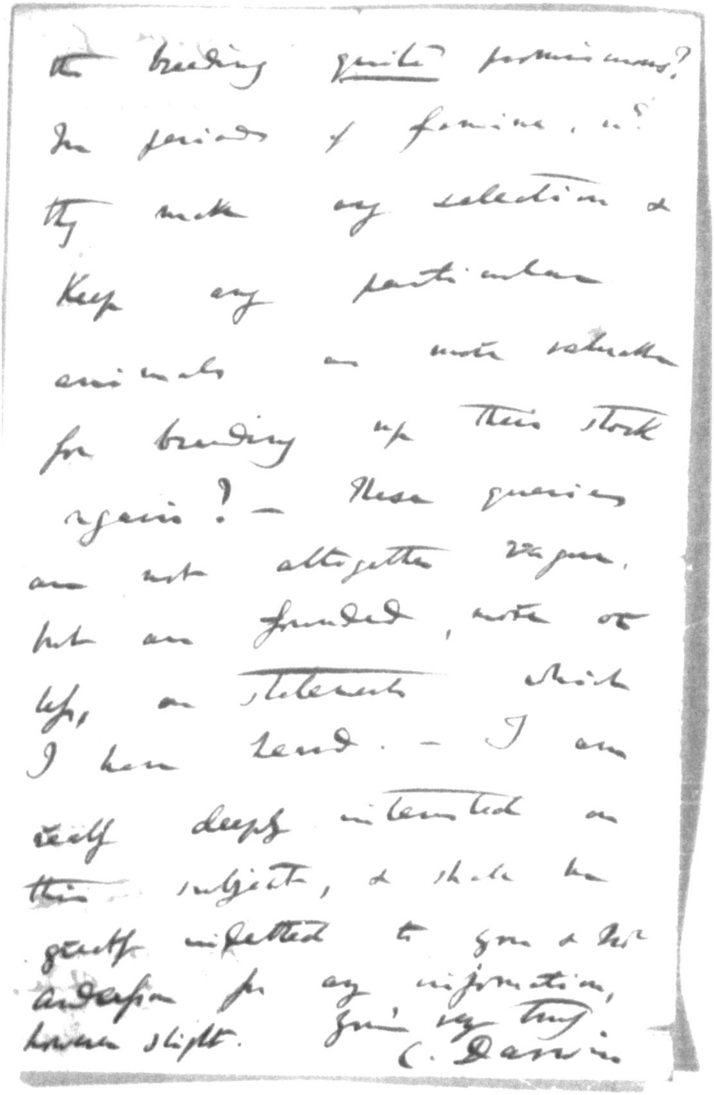

the breeding quite promiscuous?
In periods of famine, do
they make any selection &
keep any particular
animals or more valuable
for breeding up their stock
again?— These queries
are not altogether vague,
but are founded, more or
less, on statements which
I have heard.— I am
myself deeply interested on
this subject, & shall be
greatly indebted to you & Mr
Andersson for any information,
however slight. Yours very truly
C. Darwin

Figure 3.6: Charles Darwin to Francis Galton: Page 5.

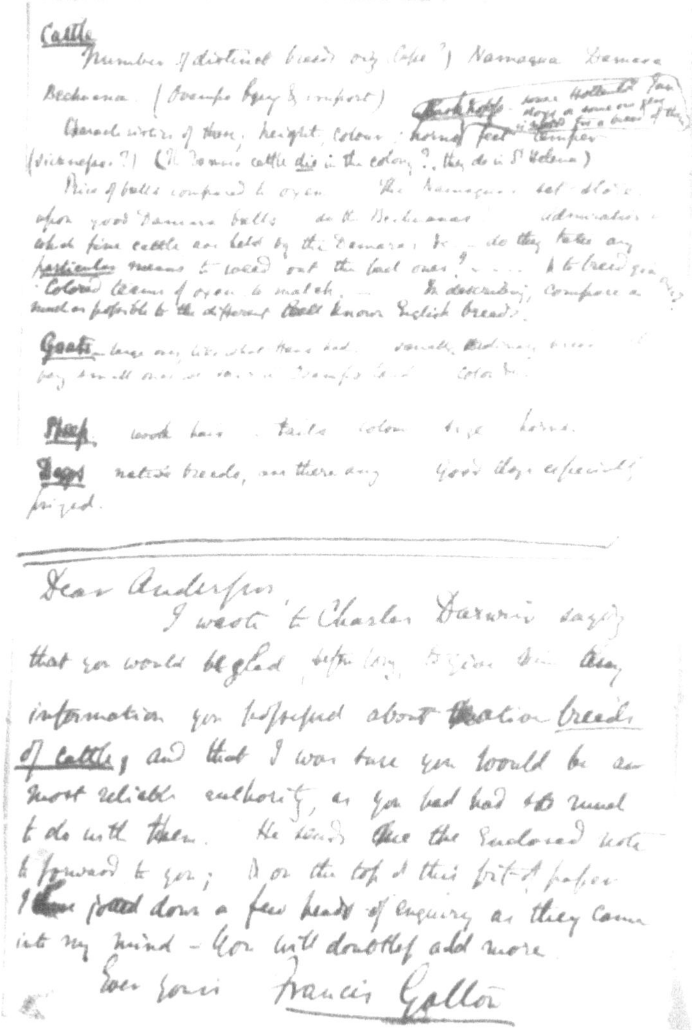

Cattle
Number of distinct breeds [illegible] Namaqua Damara Bechuana (Ovampo [illegible] & import)
Characteristics of them, height, colour, horns, feet, temper
[illegible] (Do Damara cattle die in the colony?, they do in St Helena)
Price of bulls compared to oxen. The Namaquas set store upon good Damara bulls, so do the Bechuanas [illegible] which fine cattle are held by the Damaras &c — do they take any particular means to weed out the bad ones? [illegible]
Coloured teams of oxen to match. — In describing, compare as much as possible to the different well known English breeds.
Goats — large ones, [illegible] small ones [illegible] colour &c.
Sheep wool hair, tails colour size horns.
Dogs native breeds, are there any good dogs especially prized.

Dear Andersson,
I wrote to Charles Darwin saying that you would be glad, before long, to give him any information you possessed about native breeds of cattle, and that I was sure you would be a most reliable authority, as you had had so much to do with them. He sends me the enclosed note to forward to you; & on the top of this bit of paper I have jotted down a few heads of enquiry as they came into my mind — You will doubtless add more
Ever yours Francis Galton

Figure 3.7: Francis Galton to Charles John Andersson, enclosing Darwin's letter.

Chapter 4

Negroes and the Slave Trade, 1857

TO THE EDITOR OF THE TIMES[1]

Sir,—I do not join in the belief that the African is our equal in brain or in heart; I do not think that the average negro cares for his liberty as much as an Englishman, or even as a serf-born Russian; and I believe that if we can, in any fair way, possess ourselves of his services, we have an equal right to utilize them to our advantage as the State has to drill and coerce a recruit who in a moment of intoxication has accepted the Queen's shilling, or as a shopkeeper to order about a boy whose parents had bound him over to an apprenticeship. I say an equal right, because if soldiers were abased and degraded by their profession, or if the duties of an apprentice tended to make him a worthless member of society, it would be an iniquitous exercise of tyranny to take advantage of the position of these persons to their manifest injury. But when the soldier is taught self-respect, and is made into a nobler man than he could have become if left in his village, and if the apprentice is trained into a useful member of an industrious class, there can be no just complaint of tyranny. These persons are simply treated as children by their masters, and compelled to do what they dislike for their future good and for that of society at large. Therefore, Sir, I say, with regard to these negroes, if we can by any legitimate, or even quasi-legitimate means, possess ourselves of a right to their services, and if we can insure that our mastership shall elevate them, and not degrade them, by all means work them well; but in proportion as we cannot act favourably upon them our interference becomes a curse to the Africans.

---

[1] Galton 1857a

It is often argued, "let us not be too curious about the antecedents of the negroes, who are collected by the native chiefs (of course for a 'consideration') as candidates for free emigration. Very likely they may be captured for this express purpose, but what of that? Africans are always fighting, and have no notion of personal liberty, and if the conquerors choose to sell their prisoners instead of keeping them as slaves, why should we abstain from buying?" To this I reply that the disorganization induced in the whole of African society under the temptations of the slave trade is something truly frightful. We know, Sir, in our own country the effect of laws that give a premium to crime, and your columns have often denounced them. Raise the Customhouse tariff, and honest fishermen turn into ruffianly smugglers; allow people to insure their houses for more than they are worth, and what street would be secure from fire? The very burial clubs tempted mothers to murder. The award of "head money" had effect, about which report was rife, upon our cruisers in the Malay waters. The railway mania tempted the respectable landowners of England to cheat corporate bodies without a qualm, and now the temptation of West Indian commerce seems sufficient to draw a section of Englishmen into the very vortex of an African slave trade. If, Sir, your readers will try to imagine themselves in a position where every fellow-creature is as a bank-note, which has only to be laid hold of and presented to one of the ever ready agents to be payable at sight, they will succeed in picturing to themselves the awful disorganization which must necessarily ensue—the wars for captives, the false accusations to obtain prisoners, the mutual suspicion of neighbours, and the abandonment of all steady labour for the lottery of slave catching. Most earnestly, therefore, do I deprecate an action on our part which, directly or indirectly, in the slightest degree would reintroduce a sale of negroes. The peaceful habits which have slowly been fostered among many African tribes would be swept away in a moment under the pressure of a temptation they were not strong enough to bear.

What, then, is to be our course? I cannot believe that it is impossible for an African to enter our service in the colonies without being degraded like those in America. Let the philanthropists show how we can act justly towards our blacks when we get them. Now, as to how they are to be got. I do not at all think that adequate attempts have ever been made to obtain a free African immigration. The number of recruits depend[2] on the skill of the recruiting officer. We must ingratiate ourselves more with the African tribes generally. As it is, those a few days' journey from the coast know little or nothing of us. You are doubtless aware, Sir, that the generally spread belief concerning the whites is that they buy slaves in order to carry them across sea, and there to eat them. It will require time to disabuse the native minds of these kinds of notions, but I fully believe it is to be done, and that by a consistent and judicious political action we may make our service respected, if not actually sought after; and that by watching the turn of events and taking advantage of great national suffering, such as that the Caffres are now labouring under, we may succeed in deporting vast numbers of Africans to colonies where they will do us good service, and in which we shall not have to reproach ourselves with neglecting our duty towards them.

42, Rutland-gate.

FRANCIS GALTON.

[2] sic.

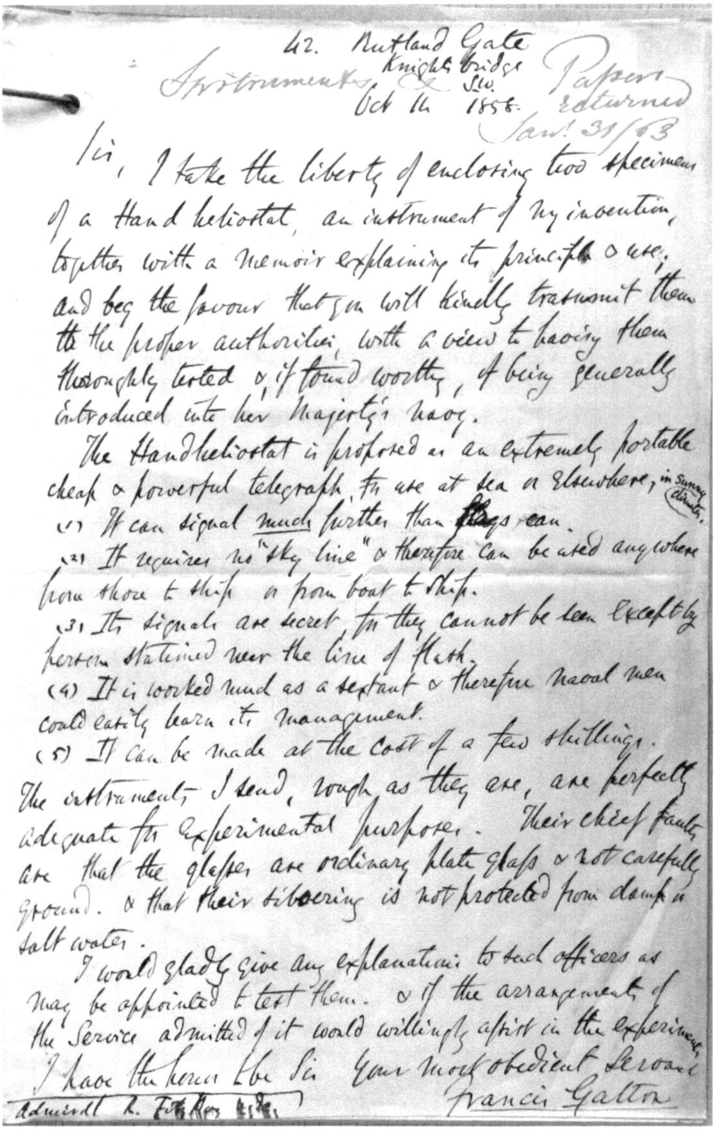

42. Rutland Gate
Knightsbridge
S.W.
Oct 14 1858.

Instruments, &c

Papers returned
Jan. 31/63

Sir, I take the liberty of enclosing two specimens of a Hand heliostat, an instrument of my invention, together with a memoir explaining its principle & use, and beg the favour that you will kindly transmit them to the proper authorities, with a view to having them thoroughly tested & if found worthy, of being generally introduced into Her Majesty's navy.

The Hand heliostat is proposed as an extremely portable cheap & powerful telegraph, for use at sea or elsewhere, in sunny climates.

(1) It can signal much further than flags can.

(2) It requires no "sky line" & therefore can be used anywhere from shore to ship or from boat to ship.

(3) Its signals are secret, for they cannot be seen except by persons stationed near the line of flash.

(4) It is worked much as a sextant & therefore naval men could easily learn its management.

(5) It can be made at the cost of a few shillings.

The instruments I send, rough as they are, are perfectly adequate for experimental purposes. Their chief faults are that the glasses are ordinary plate glass & not carefully ground, & that their silvering is not protected from damp or salt water.

I would gladly give any explanations to such officers as may be appointed to test them, & if the arrangements of the Service admitted of it would willingly assist in the experiments.

I have the honor to be Sir Your most obedient Servant
Francis Galton

Admiral R. FitzRoy &c.

Figure 4.1: Francis Galton to Admiral Robert Fitzroy, 1858-10-14, concerning his hand-heliostat.

Chapter 5

Supplementary Materials to *The Reader* magazine

5.1 The British Association Debate of 1860.

5.1.1 Oxford Chronicle.

We are favoured by a correspondent[1] with the following succinct report of the animated and interesting discussion which took place at the Library of the New Museum on Saturday, the 7th inst., upon Professor Draper's[2] paper bearing upon Mr. Darwin's Theory of the Origin of Species, which was noticeable for a sharp passage of arms between Prof. Huxley and the Bishop of Oxford. Long before the time fixed (12 o'clock) an immense audience had assembled, occupying every available space, and many were unable to gain admittance.

SECTION D.—"On the Intellectual Development of Europe, considered with reference to the Views of Mr. Darwin and others, that the Progression of Organisms is determined by Law." By Dr. Draper, Professor of Physiology in the University of New York.

The purpose of this paper was to show that the advance of human civilization is not accidental or fortuitous, but is in obedience to immutable law. It exhibited a research and philosophical power well deserving the careful and respectful hearing which it obtained. [Draper went on to argue that the lives of nations pass through stages of development like those of individuals.] Referring to his work on physiology,

1 Saturday 21 July 1860, page 3. Expands on a shorter account given on Saturday 14 July 1860, 5.

2 John William Draper (1811–1882).

published in 1856, he then gave his views in support of the doctrine of the transmutation of species; the transitorial forms of the animal and also the human type; the production of new ethnical elements or nations, and the laws of their origin, duration, and death. The Professor, on resuming his seat, was loudly cheered.

Professor HUXLEY, being called upon by the chairman, declined entering into the subject, alleging the undesirability of contesting a scientific subject involving nice shades of idea before a general audience, who could not be supposed to judge upon its merits.

The discussion was then commenced by the Rev. R. GRESWELL, who denied that any parallel could be drawn between the intellectual progress of man, and the physical development of the lower animals. He disputed Professor Draper's references to the history of Greece. Was it not a fact that its masterpieces in literature, the Iliad and the Odyssey, were produced during its national infancy. (Hear, hear.) The theory of intellectual development advanced, he considered, was directly opposed to the known facts in the history of man. (Applause.)

Sir B. BRODIE, who on rising was very warmly received, said he could not subscribe to the hypothesis of Mr. Darwin. Where was the demonstration that his primaeval germ had existed? (Hear, hear.) Man had a power of self-consciousness — a principle differing from anything found in the material world, and he did not see how this could originate in lower organisms. Moreover this power of man, being identical with the Divine intelligence, to suppose that it could originate with matter, involved the absurdity of supposing the source of Divine Power was dependent on the arrangement of matter. (Loud cries of "hear, hear," and much applause.)

The BISHOP OF OXFORD, on rising, was loudly cheered. Having briefly noticed Professor Draper's arguments, and ridiculed his comparison of the marble and table, he proceeded to say he had given the theory advanced by Mr. Darwin his most careful and anxious consideration. The conclusion he had come to was, that when tried by the principles of inductive science, philosophy, or logic, it entirely broke down. (Cheers.) And to come to facts, he maintained that those brought forward utterly failed to prove his theory. The permanence of specific forms was a fact confirmed by all observation, the few exceptions that existed being confined to a few cases in certain species of plants. Take, for instance, the remains of the animals, plants, and man, found in those earliest records of the human race—the Egyptian catacombs. Now, anatomists tell us that even in mummies, 4,000 years old, there is not the slightest physiological difference as compared with the race now—(hear, hear)—and so it was with animals and plants. All spoke of their identity with existing forms, and of the irresistible tendency of organised beings to assume an unalterable character. (Applause.) Glancing at Professor Huxley's remarks, on the previous day, in a discussion with Professor Owen, the Bishop facetiously asked if he had any particular predilection for a monkey ancestry, and, if so, on which side—whether he would prefer an ape for his grandfather, and a woman for his grandmother, or a man for his grandfather, and an ape for his grandmother. (Much laughter.) But to treat the subject seriously. (Hear, hear.) The line between man and the lower animals was distinct. There was no tendency on the part of the lower animals to become the self-conscious intelligent being, man; or, in man, to degenerate and lose the high characteristics of mind and intelligence. All experiments failed to show any tendency in one animal to assume the form of the other. Even in the great case of pigeons, quoted by Mr. Darwin, he admitted that no sooner were these animals set free, than they returned to their primitive type. (Hear, hear.) Everywhere sterility attended hybridism, as was seen in the closely allied forms of the horse and the ass. Viewing the matter in another aspect, he did consider it a most degrading assumption—(hear, hear)—that man, who, in many respects, partook of the highest attributes of God—(hear, hear)—was a mere development of the lowest forms of creation. (Applause.) He could scarcely trust himself to speak upon the subject, so indignant did he feel at the idea. He did not desire timidity in scientific investigation. (Hear, Hear.) Religion had nothing to fear. (Hear, hear) But what he did protest against was the hasty adoption of unsound hypotheses and unproved assertions for the weighty realities of scientific truth. (Applause.) He did not believe that science and revelation were inimical to each other, but that what appeared irreconcilable in the present state of scientific knowledge would in the fulness of time be made manifest, and redound to the triumph of both. (Prolonged cheering.)

Professor HUXLEY followed. In reply to the Bishop's query he said that if the alternative were given him of being descended from a man conspicuous for his talents and eloquence, but who misused his gifts to ridicule the laborious investigators of science and obscure the light of scientific truth, or from the humble origin alluded to, he would far rather choose the latter than the former. (Oh, oh, and laughter and cheering.) He then defended Mr. Darwin's theory from the charge of being a mere hypothesis, and said it was an explanation of phenomena in natural history holding the same relation as the undulating theory to the phenomena of light. Did any one object to that theory because an undulation of light had never been arrested and measured? His charge against Mr. Darwin's opponents was that they did not attempt to bring forward any important fact against his theory. That theory was an explanation of facts the result of laborious research, and abounded in new facts bearing upon it. Without asserting that every part of the theory had been confirmed, he maintained that it was the best explanation of the origin of species which had yet been offered—(hear, hear)—and he did protest against this subject being dealt with by amateurs in science, and made the occasion of appeals to passion and feeling. (Applause.) With regard to the psychological distinction between man and animals, it must be remembered man himself was once a nomad[3]—a mere atom of matter—and who could say at what moment of his development he became consciously intelligent. (Hear, hear.) The question was not so much one of a transmutation or transition of species as of the production of forms which became permanent. In the course of an argument to support this position, he instanced the short-legged sheep of America, which were not produced gradually, but originated in the birth of an original parent of the whole stock, which had been kept up by a rigid system of artificial selection. (The Professor, on resuming his seat, was loudly applauded.)

The BISHOP OF OXFORD again rose and was received with cheers and laughter. He said he regretted that Professor Huxley had taken umbrage at what he had said. He did not know that he had said anything which could possibly give offence to Mr Darwin's greatest friends, and as for his query to Professor Huxley he had been tempted to it by the merriment of the audience, and it was merely a passing allusion. He ridiculed Professor Huxley's appeal to authority in connection with his remarks on amateurs in science. On which side lay

[3] Possibly a mistake for monad.

the authority. Sir B. Brodie, Professor Owen, and other eminent men were opposed to it, and how the Professor could talk as he had done about authority he did not know. (Laughter and cheers.) The Bishop then noticed the Professor's concluding remarks, denying the cogency of the illustrations, and after experiencing some interruptions in his scientific dicta, sat down amid loud cheers.

Professor HUXLEY rose in answer to calls for him, and said he was sure the Bishop could have no desire to mislead, but he thought he had misapprehended his remarks upon authority. What he had deprecated was authority like the Bishop's, authority derived from a reputation acquired in another sphere. (Hear, hear, and laughter.)

Dr. BIRD,[4] who announced himself as a statistician, asserted that he could prove Mr. Darwin's theory was unsound by statistics. (Murmurs.) By Mr. Darwin's theory he could prove anything. (Question.) The learned Doctor's remarks were cut short by the impatience of the audience.

Admiral FITZROY stated, as an old friend of Mr. Darwin's, he deeply regretted the views he had put forth (loud cries of question). He denied Professor Huxley's statement that Mr. Darwin's work was a logical arrangement of facts, and was proceeding to theological considerations when the interruptions became so noisy that the chairman requested the gallant admiral to sit down.

Professor BEALE said, although he was quite unable to decide the question on one side or the other, he would just notice for Professor Huxley's consideration some of the difficulties which Mr. Darwin's theory had to deal with, and among others referred especially to those vital tendencies of allied species which seemed independent of all external agents. (Cheers.)

Mr. LUBBOCK reiterated Professor Huxley's assertion that Mr. Darwin's book was the most logical and powerful arrangement of facts that had ever been given upon the subject, and expressed his willingness to accept his hypothesis in the absence of any better. He expressed surprise at the Bishop of Oxford's reference to the Egyptian mummies, since to the naturalist as to the geologist, time was not an essential element in these changes. Time alone produced no change. (Cheers.)

The Rev. R. GRESWELL rose indignantly to oppose the

[4] James Bird (1797–1864).

assertion of Professor Huxley, that man was originally an atom of matter. He denied that he was ever an atom of matter. (Oh, oh, and loud laughter, amid which the rev. gentleman sat down.)

Dr. HOOKER (the celebrated botanist), being called upon by the President to state his views of the botanical aspect of the question, observed, that the Bishop of Oxford having asserted that all men of science were hostile to Mr. Darwin's hypothesis,—whereas he himself was favourable to it,—he could not presume to address the audience as a scientific authority. As, however, he had been asked for his opinion, he would briefly give it. In the first place, his Lordship, in his eloquent address, had, as it appeared to him, completely misunderstood Mr. Darwin's hypothesis: his Lordship intimated that this maintained the doctrine of the transmutation of existing species one into another, and had confounded this with that of the successive development of species by variation with natural selection. The first of these doctrines was so wholly opposed to the facts, reasonings, and results of Mr. Darwin's work, that he could not conceive how any one who had read it could make such a mistake,—the whole book, indeed, being a protest against that doctrine. Then, again, with regard to the general phenomena of species, he understood his Lordship to affirm that these did not present characters that should lead careful and philosophical naturalists to favour Mr. Darwin's views. To this assertion Dr. Hooker's experience of the Vegetable Kingdom was diametrically opposed. He considered that at least one half of the known kind of plants were disposable in groups, of which the species were connected by varying characters common to all in that group, and sensibly differing in some individuals only of each species; so much so that, if each group be likened to a cobweb, and one species be supposed to stand in the centre of that web, its varying characters might be compared to the radiating and concentric threads, when the other species would be represented by the points of union of these; in short, that the general characteristics of orders, genera, and species amongst plants differed in degrees only from those of varieties, and afforded the strongest countenance to Mr. Darwin's hypothesis. As regards his own acceptation of Mr. Darwin's views, he expressly disavowed having adopted them as a creed. He knew no creeds in scientific matters. He had early begun the study of natural science under the idea that species were original creations; and it should be steadily kept in view that this was merely another hypothesis, which in the abstract was neither more nor less entitled to acceptance than Mr. Darwin's: neither was, in the present state of science, capable of demonstration, and each must be tested by its power of explaining the mutual dependence of the phenomena of life. For many years he had held the old hypothesis, having no better established to adopt, though the progress of botany had, in the interim, developed no new facts that favoured it, but a host of most suggestive objections to it. On the other hand, having fifteen years ago been privately made acquainted with Mr. Darwin's views, he had during that period applied these to botanical investigations of all kinds in the most distant parts of the globe, as well as to the study of some of the largest and most different Floras at home. Now, then, that Mr. Darwin had published it, he had no hesitation in publicly adopting his hypothesis, as that which offers by far the most probable explanation of all the phenomena presented by the classification, distribution, structure, and development of plants in a state of nature and under cultivation; and he should, therefore, continue to use his hypothesis as the best weapon for future research, holding himself ready to lay it down should a better be forthcoming, or should the now abandoned doctrine of original creations regain all it had lost in his experience. The learned gentleman, during his address, was repeatedly cheered, and resumed his seat amid loud applause. The President then declared the section adjourned.

5.1.2 The Atlas.

LITERATURE.[5] The closing meeting of the British Association this year took place on Wednesday in the Sheldonian theatre. Oxford has been the theatre of their antiquarian researches; but the "lions" of this antique town have been so often "done," that little original interest has been created by the recent gathering of our savants. In fact, the whole proceedings have gone off with a tameness and insipidity not a little discouraging. The only events by which the session will probably be remembered are the discussions which took place on Friday, Saturday, and Tuesday, when the *odium theologicum* had full sway, and when (which is still more remarkable), a powerful support was given to certain new theories

[5] Saturday 07 July 1860, 10.

regarded as by no means orthodox by the "unco gude." On the Friday morning, there was a strong mustering in the Economical Section, as it was known that a formidable attack would be made on certain views put forward by Dr. Whewell in a prefatory notice to the works of the late Mr. Jones. The intellectual athlete who entered the lists was Mr. Fawcett, a young Cambridge scholar, afflicted with blindness, we believe, who addressed himself to the task with peculiar energy and logical acumen. Dr. Whewell made an elaborate reply, and was applauded by his friends; but the general opinion entertained by the auditory was that his adversary had the best of the attack. Saturday, however, was the great day—the day of days. It had oozed out that Mr. Darwin's theory was to be assaulted by Samuel, Bishop of Oxford, and eager were the academical multitude to catch a glimpse of this grand feat of arms. Some home-thrusts and skilful parries were expected, to give greater zest to the contest. The time arrived, and the right reverend prelate arose in his might. He denounced in eloquent and emphatic language Mr. Darwin's theory, and was apparently carrying all before him, when, unfortunately, his evil genius tempted him to be personal. He endeavoured to create a feeling of the ridiculous by suggesting that a disciple of Darwin's, like Mr. Huxley, would not shrink from the conclusion, inevitably drawn from the premisses laid by the physiologist, that his father might have been an ape. Achilles had one vulnerable point; so, it appears, had the Bishop, and that was it; for Mr. Huxley, rising, retorted in reply, that, "if the alternative was offered him, on the one hand, to claim as his father an ape, and on the other hand, to become a man of brilliant attainments, skilled in the fence of words, who should devote his faculties to the misrepresentation of facts, and to throw ridicule on a scientific discussion—if this alternative were offered, he would accept, he exclaimed, "the poor despised ape." The debate was closed by an impressive discourse from Dr. Hooker, the eminent botanist, who strenuously vindicated the views of Darwin, and was supported by the general approbation of those present. On Tuesday, the contest was transferred to the Ethnological section, when Dr. Hunter maintained a theory in favour of the great antiquity of man. This again excited the theological spirit of a crowded meeting. Mr. Chambers seconded the views of Dr. Hunter, and brought forward, as a strong argument in support of the theory, the flint-arrow-heads lately discovered in France. The opponents of these views in vain attempted to shake his geological reasoning. Thus, a triple victory has been obtained for science at the Oxford gallery of the British Association; but, beyond these discussions, the meetings have passed off with very little éclat. The meeting of 1861, it was announced by Professor Phillips, will take place at Manchester, and most probably in the month of August.

5.1.3 Morning Chronicle.

... The Bishop of Oxford's question is a good one;[6] Mr. Huxley's answer is a good one—that he would rather both his great great great ancestors were quadrumanous, and state the enormous truth, than deny the truth or palter with it. ... Piccadilly, July 4. HARPOCRATES

5.1.4 Cheltenham Chronicle.

Professor Huxley and the Bishop of Oxford.[7]—In the discussion on Mr. Darwin's theory, the British Association, Professor Huxley had remarked that people had nothing to fear even should it be shown that apes were their ancestors. The Bishop of Oxford on a subsequent day, taking the hint, asked the Professor whether he would prefer a monkey for his grandfather or his grandmother. To this the man of science sharply retorted that he would much rather have a monkey for his grandfather than man who could indulge in jokes on such a subject.

5.2 T. H. Huxley on Science and Church Policy.

SCIENCE AND "CHURCH POLICY."[8]

6 Monday 09 July 1860, 7.

7 Tuesday 17 July 1860, 7. There were similar reports in Aris's Birmingham Gazette, Saturday 14 July 1860, 10; Bucks Advertiser & Aylesbury News, Saturday 28 July 1860, 2–3; Morning Advertiser, Tuesday 31 July 1860, 3–4; Worcestershire Chronicle, Wednesday 25 July 1860, 4; The Suffolk Chronicle; or Weekly General Advertiser & County Express, Saturday 28 July 1860, 2; Reading Mercury, Saturday 07 July 1860, 10; Cheltenham Chronicle, Tuesday 17 July 1860, 7; Birmingham Journal, Saturday 28 July 1860, 7; Glasgow Herald, Wednesday 04 July 1860, 4 (where Huxley is named 'Wrexley' and other details are garbled).

8 *The Reader.*

THE attitude towards scientific thought adopted by our ecclesiastical dignitaries, by our statesmen, and by those political intriguers who embody the ideal set forth in "Vivian Grey" and "Coningsby," and who, though of little faith enough in most things, yet believe in the efficacy of a "church cry" for electioneering purposes, is worthy of serious attention.

Our leading statesmen—unless the Premier be an exception — neither know nor care anything about Science. Not long ago one of them, with characteristic frankness, told us as much, and hardly cared to throw out the excuse that "he was born in the pre-scientific age." Every hare-brained speculator, or enterprising charlatan, knows full well that the minister, whose patronage he solicits, would not be deterred from according it by the mere circumstance that his scheme bears upon its face some glaring contradiction to the law of gravitation, or the doctrine of the persistence of force. The minister is no less conscious of the fact, and, to protect himself, treats quack and genius alike to the practical negative of official circumlocution—"under consideration."

Turn to the Houses of Lords and Commons, and we find the most prominent, and some of the most accomplished, of their members agreed in this, if in nothing else, that the notion of introducing some branches of physical science into the schooling of an English gentleman is a mere mischievous chimera.

They urge, justly and wisely enough, that the value of education lies in its training, rather than in its teaching; and therefore (they naively ask) what possible mental discipline is to be obtained from learning to use the lenses and the prisms, the wires and the crucibles, the forceps and the scalpels of the man of science? In the same spirit we have seen enlightened public writers settle the whole question of the advisability, or inadvisability, of scientific teaching by the crushing inquiry, " How will an Eton boy be the better for knowing how to make a pump? Doubtless it is a good thing to know how to make a pump; but it is also a good thing to know how to make shoes; and yet you do not propose to introduce shoemaking as a branch of liberal education." And, as this argument constituted the climax of a well-polished essay, it is clear that the essayist really did think it relevant, and imagined that the rest of the world would appreciate it.

When enlightened statesmen and thoughtful public writers hold such views, it is not wonderful to find Mr. Disraeli adopting a grotesque imitation of them. For Mr. Disraeli is an actor of astonishing power and versatility. You read his novels and wonder that something so like genius can be achieved by mere versatile cleverness; you listen with equal surprise to that instinctively dexterous rhetoric—so like oratory—which makes the most threadbare speech, or even a secondhand one, look better than new. And, on perusing his latest effort, just published for the benefit of a wider world than the Society for the Augmentation of Small Livings, marvel rises to admiration at the excellently simulated earnestness with which the rhapsodist of that misty sentimentality, the Asian mystery, reproves "picturesque eloquence" and "the lucubrations of nebulous professors;" while admiration soars to ecstasy when the great mime " presents," as Bottom would say, the man of science—prattles of the "bracing effects of original research," and warns us against " secondhand knowledge."

So well was it all done that, if the printed version of the speech is to be trusted, nobody laughed. Even the Right Rev. Prelate in the chair, among whose many graces the keen enjoyment of a joke is not the least, preserved his gravity, and listened patiently, as the fervid speaker went on to explain the relations of Science to the Church.

In substance, what Mr. Disraeli tells us on this head is, that philosophers are all very well in their proper places—making steam-engines and telegraphs and railroads, and adding "to the convenience of life and to the comfort of them." But let there be no more "tattle man." Let them stick to this, and Mr. Disraeli is kindly prepared "to do full justice to about Science"[9] in any other capacity, or his wrath and scorn will know no bounds.

Of course, Mr. Disraeli's real, or assumed, opinions on questions of this kind are of no importance in themselves. We draw attention to them because it really is a wonderful phenomenon that one, so practised in adapting himself to the humour of an assembly, should have felt that patronizing Science for its froth and scum—its so-called practical results —and scorning its essence and the foundation of its human worth, was, in the year of grace 1864, a safe card to play before bishops, dons, and members of Parliament assembled in conclave in one of the great seats of learning of this country.

When the laity, who have no professional prepossessions to overcome, deal thus strangely with Science, we must not look for better things from ecclesiastics. Amiable and lib-

[9] sic.

eral prelates, unable to discern any true savour of diabolic agency in the uniformly beneficent and ennobling working of scientific thought, honestly wishing it well, and yet genuinely afraid of it, plaintively ask why it cannot be content. "We have thrown overboard our Astronomy, our Physics, our Geology, our Chronology and Ethnology, one after another; we have almost made up our minds to sacrifice three-fourths of our History, and to go shares with you in Ethics; we even admit—oh, philosophers !—that "a religion can have no claim to be accepted as coming from God which contradicts or overlooks the grand principles of immutable morality" (which can be determined only by a strict application of scientific method)—and yet you are not content. What, then, do you mean to leave us?" is the substance of an admirable address recently delivered at Edinburgh.

Suave and illiberal prelates, on the other hand, shrewder—though, it may turn out, not wiser—than the excellent Bishop of London, disbelieve in the possibility of compromise, and, apparently sympathizing with the spirit, if not ready to abide by the letter, of the last encyclical epistle of the Bishop of Bishops, think to stay the spirit of the age by copious malediction of its products, tendencies, and representative men.

Such seems to us to be a fair view of the present attitude of those who govern the State and the Church towards Science and her methods. But what of the other side, how do the men of science view the question? So far as we can judge, they seem in no humour to take the proffered position. Science exhibits no immediate intention of signing a treaty of peace with her old opponent, nor of being content with anything short of absolute victory and uncontrolled domination over the whole realm of the intellect. Her champions ask why they should falter? Which of the memorable battles that have been fought have they lost? When have they ever retreated from ground they have once occupied, or surrendered a duly fortified stronghold? Every invading host has its guerillas and skirmishers, who sometimes advance into untenable positions and are beaten back; or burn, plunder, and ravish to the utter disgrace of themselves and their cause, and, justly enough, fall into the hands of the provost-marshal. But their rout is no retreat of the main army, and their discomfiture is a source of strength rather than of weakness.

And so, if, as the orator of Oxford tells us, the scoffing light-horsemen of the eighteenth century have been beaten off, and the old traditions have emerged from the smoke not much hurt, is that a repulse of Science? Surely not; for the broken squadrons of Voltairean cossacks fly only to disclose the heads of solid columns of warriors, disciplined in long and successful struggles with nature, firm of purpose, and armed with weapons which are unpoisoned but slay effectually. And so strangely are circumstances altered, that it seems to many as if the Voltaireans had taken shelter among the enemy's ranks; while the spirit of faith and reverence, and the will to die rather than to lie, had left the theological and entered the scientific camp.

The philosophy of the present day, indeed, is neither scoffing, nor presumptuous, nor destructive. Since the world began, there never has been so deep a reverence for truth, so keen a sense of the fallibility and limitation of the intellect of man, so earnest a desire to build up some theory of this wonderful universe that cannot be shaken by the questioning of a child, so profound a yearning

"Im Guten, Ganzen, Wahren, resolut zu leben."

as among the scientific workers of this age and generation.

Never has there been so clear an appreciation of the unity of all phenomena, and hence of the absurdity of both materialism and spiritualism. Never has the consequent necessary identity of the methods by which truth of all kinds must be attained been so clearly obvious as it is now. And, therefore, never has the attempt to set bounds to scientific inquiry and to the extension of scientific method, into every subject concerning which a proposition can be framed, proclaimed itself at once so fatuous and so impotent as now.

Religion has her unshakeable throne in those deeps of man's nature which lie around and below the intellect, but not in it. But Theology is a simple branch of Science, or it is nought; and that "Church Policy" which sets it up against Science is about as reasonable, as would be the advocacy of the claims of the rule of three[10] to superior authority over arithmetic in general.

[10] [Ed.] In an equation involving four variables, the value of the fourth can be deduced from the other three if they are known.

Chapter 6

Review by Alfred Russel Wallace of *Hereditary Genius*

HEREDITARY GENIUS[1]

Hereditary Genius, an Inquiry into its Laws and Consequences. By Francis Galton, F.R.S., &c. (Macmillan & Co.)

IN this book Mr. Galton proposes to show that a man's natural abilities are derived by inheritance, under exactly the same limitations as are the form and physical features of the whole organic world. Many who read it without the care and attention it requires and deserves, will admit that it is ingenious, but declare that the question is incapable of proof. Such a verdict will, however, by no means do justice to Mr. Galton's argument, which we shall endeavour to set forth as succinctly as possible. He first discusses the classification of men by "reputation," and from a study of biographical dictionaries and obituaries for certain years taken at wide intervals, arrives at the conclusion that not more than 250 men in each million, or 1 in 4,000, can be termed "eminent"; and he shows what a small proportion that is, by the well-known fact that there are never so many as 4,000 stars visible to the naked eye at once, and that we feel it to be an extraordinary distinction in a star to be the brightest in the sky. These "eminent" men are the lowest class he deals with. The more illustrious names are as one in a million or one in many millions; but unless a man is so much above the average that there is only one like him in every 4,000, he is not admitted into the ranks of the eminent men on whom Mr. Galton founds his deductions.

He next discusses the classification of men according to

[1] Wallace 1870.

their natural gifts. He shows first, that each man has a certain defined limit to his mental as well as to his physical powers, and that this limit is in most cases soon discovered and reached. He next shows the enormous difference that exists between mediocre and high class men, by the evidence of examination papers; the senior wrangler at Cambridge, for example, often getting thirty times as many marks as the lowest wrangler, who must himself be a man very far above the average. Statistics show, that the number of imbeciles and idiots are about the same per thousand as the eminent men. He then applies Quetelet's "law of deviation from an average" (which will be new to many of his readers), and deduces from it, that if men are divided into sixteen equal grades of ability, eight above and eight below the average, the six mediocre classes will comprise nineteen-twentieths of the whole; while it will be only the sixth, seventh, and eighth above the average who will rank as eminent and illustrious men, and form about one in four thousand of the adult male population.

The next chapter relates to the important question on which, indeed, the possibility of any solution of the problem depends, of whether "reputation" is a fair test of "ability." The subject is very ably discussed, and it is, I think, proved, that notwithstanding all the counteracting influences which may repress genius on one side, or give undue advantage to mediocrity on the other, the amount of ability requisite to make a man truly "eminent" will, in the great majority of cases, make itself felt, and obtain a just appreciation. But if this be the case, the question of whether "hereditary genius" exists is settled. For if it does not, then, the proportion of mediocre to eminent men being 4,000 to 1, we ought to find that only 1 in 4,000 of the relations of eminent men are themselves eminent. Every case of two brothers, or of father and son, being equally talented, becomes an extraordinary coincidence; and the mass of evidence adduced by Mr. Galton in the body of his work, proves that there are more than a hundred times as many relations of eminent men who are themselves eminent, than the average would require.

Turning now to the concluding chapters of the book, we meet with some of the most startling and suggestive ideas to be found in any modern work. The law of deviation from an average enables us to determine the general intellectual status of any nation, if we are able to estimate the ability of its most eminent men, and know approximately the amount of the population. We have these data in the case of ancient Attica; and Mr. Galton arrives at the conclusion, that the Athenians of the age of Pericles were, on the lowest possible estimate, nearly two whole grades of ability higher than we are. With all our boasted civilisation, and the vast social and scientific problems with which we have to grapple; with all our world-wide interests, our noble literature, and accumulated wealth; the intellectual status of the most civilised modern nation is actually lower than it was more than two thousand years ago! Well may Mr. Galton maintain that it is most essential to the well-being of future generations that the average standard of ability of the present time should be raised. Not less striking is his exposition of the effects of prudential restraints on marriage, on the general character of a nation. If one class of people, as a rule, marry early, and another class marry late in life, the former have a double advantage, both in having on the average larger families, and in producing more generations in each century. But, by the supposition, it is the imprudent who gain this advantage over the prudent; and Mr. Galton therefore denounces the doctrine of Malthus, that marriage should be delayed till a family can be supported, unless the rule could be imposed on all alike. I hardly think that this argument is sound, and I doubt if the imprudent who make early marriages do, in the long run, increase more rapidly than the prudent who marry late. Increase of population depends less upon the number of children born, than on those which reach manhood; and I believe that the prudent man who has acquired some wealth and wisdom before he marries, will give to the world more healthy men and women, than the ignorant and imprudent youth, who marries a girl as ignorant and imprudent as himself. It is also to be remembered that the men who marry late often marry young wives, and have as good a chance of large families as the imprudent.

Mr. Galton traces the long-continued darkness of the Middle Ages, and our present low intellectual and moral status, to the practice of celibacy and to religious persecution. Whenever men and women were possessed of gentle natures, that fitted them for deeds of charity, for literature, or for art, the social condition of the times was such that they had no refuge but in the bosom of the Church; and the Church exacted celibacy. Those gentle natures left no offspring; and thus was the race of our forefathers morally deteriorated. The Church acted as if she had aimed at selecting the rudest portion of the community for the parents of future generations; and the rules as to fellowships at our Universities are a relic of this barbarous custom, being bribes to men of exceptional ability not to marry. Religious persecution acted in the same way. The most fearless, truth-seeking, and intelligent were year by year incarcerated in dungeons or burned at the stake; so that, by this twofold selection, human nature was brutalised and demoralised, and we still feel its hateful effects in the long-continued antagonism to the essential requirements of an advancing civilisation. These concluding chapters stamp Mr. Galton as an original thinker, as well as a forcible and eloquent writer; and his book will take rank as an important and valuable addition to the science of human nature.

Alfred R. Wallace

Chapter 7

Statistical Inquiries into the Efficacy of Prayer[1]

An eminent authority has recently published a challenge to test the efficacy of prayer by actual experiment. I have been induced, through reading this, to prepare the following memoir for publication, nearly the whole of which I wrote and laid by many years ago, after completing a large collection of data, which I had undertaken for the satisfaction of my own conscience. The efficacy of prayer seems to me a simple, as it is a perfectly appropriate and legitimate subject of scientific inquiry. Whether prayer is efficacious or not, in any given sense, is a matter of fact on which each man must form an opinion for himself. His decision will be based upon data more or less justly handled, according to his education and habits. An unscientific reasoner will be guided by a confused recollection of crude experience. A scientific reasoner will scrutinise each separate experience before he admits it as evidence, and will compare all the cases he has selected on a methodical system.

The doctrine commonly preached by the clergy is well expressed in the most recent, and by far the most temperate and learned of theological encyclopaedias, namely, "Smith's Dictionary of the Bible." The article on "Prayer," written by the Rev. Dr. Barry, states as follows: "Its real objective efficacy is both implied and expressed [in Scripture] in the plainest terms We are encouraged to ask special blessings, both spiritual and temporal, in hopes that thus, and thus only, we may obtain them. .... It would seem the intention of Holy Scripture to encourage all prayer, more especially intercession, in all relations and for all righteous objects." Dr. Hook, the present

[1] Galton 1872b.

Dean of Chichester, states in his "Church Dictionary," under "Prayer," that "the general providence of God acts through what are called the laws of nature. By his particular providence God interferes with those laws, and he has promised to interfere in behalf of those who pray in the name of Jesus. We may take it as a general rule that we may pray for that for which we may lawfully labour, and for that only." The phrases of our Church service amply countenance this view; and if we look to the practice of the opposed sections of the religious world, we find them consistent in maintaining it. The so-called "Low Church" notoriously places absolute belief in special providences accorded to pious prayer. This is testified by the biographies of its members, the journals of its missionaries, and the "united prayer-meetings" of the present day. The Roman Catholics offer religious vows to avert danger; they make pilgrimages to shrines; they hang votive offerings and pictorial representations, sometimes by thousands, in their churches, of fatal accidents averted by the manifest interference of a solicited saint.

A prima facie argument in favour of the efficacy of prayer is therefore to be drawn from the very general use of it. The greater part of mankind, during all the historic ages, have been accustomed to pray for temporal advantages. How vain, it may be urged, must be the reasoning that ventures to oppose this mighty consensus of belief! Not so. The argument of universality either proves too much, or else it is suicidal. It either compels us to admit that the prayers of Pagans, of Fetish worshippers, and of Buddhists who turn praying-wheels, are recompensed in the same way as those of orthodox believers; or else the general consensus proves that it has no better foundation than the universal tendency of man to gross credulity. The collapse of the argument of universality leaves us solely concerned with a simple statistical question—are prayers answered, or are they not? There are two lines of research, by either of which we may pursue this inquiry. The one that promises the most trustworthy results is to examine large classes of cases, and to be guided by broad averages; the other, which I will not employ in these pages, is to deal with isolated instances. An author who made much use of the latter method might reasonably suspect his own judgment —he would certainly run the risk of being suspected by others—in choosing one-sided examples.

The principles are broad and simple upon which our inquiry into the efficacy of prayer must be established. We must gather cases for statistical comparison, in which the same object is keenly pursued by two classes similar in their physical, but opposite in their spiritual state; the one class being prayerful, the other materialistic. Prudent pious people must be compared with prudent materialistic people, and not with the imprudent nor the vicious. Secondly, we have no regard, in this inquiry, to the course by which the answer to prayers may be supposed to operate. We simply look to the final result—whether those who pray attain their objects more frequently than those who do not pray, but who live in all other respects under similar conditions. Let us now apply our principles to different cases.

A rapid recovery from disease may be conceived to depend on many causes besides the reparative power of the patient's constitution. A miraculous quelling of the disease may be one of these causes; another is the skill of the physician, or of the nurse; another is the care that the patient takes of himself. In our inquiry, whether prayerful people recover more rapidly than others under similar circumstances, we need not complicate the question by endeavouring to learn the channel through which the patient's prayer may have reached its fulfilment. It is foreign to our present purpose to ask if there be any signs of a miraculous quelling of the disease, or if, through the grace of God, the physician had showed unusual wisdom, or the nurse or the patient unusual discretion. We simply look to the main issue—do sick persons who pray, or are prayed for, recover on the average more rapidly than others?

It appears that, in all countries and in all creeds, the priests urge the patient to pray for his own recovery, and the patient's friends to aid him with their prayers; but that the doctors make no account whatever of their spiritual agencies, unless the office of priest and medical man be combined in the same individual. The medical works of modem Europe teem with records of individual illnesses and of broad averages of disease, but I have been able to discover hardly any instance in which a medical man of any repute has attributed recovery to the influence of prayer. There is not a single instance, to my knowledge, in which papers read before statistical societies have recognised the agency of prayer either on disease or on anything else. The universal habit of the scientific world to ignore the agency of prayer is a very important fact. To fully appreciate the "eloquence of the silence" of medical men, we must bear in mind the care with which they endeavour to assign a sanatory value to every influence. Had prayers for the sick any notable effect, it is incredible but that the doctors, who are always on the watch for such things, should have observed it, and added their influence to that of the priests towards obtaining them for every sick man. If they abstain from doing so, it is not because their attention has never been awakened to the possible efficacy of prayer, but, on the contrary, that although they have heard it insisted on from childhood upwards, they are unable to detect its influence. Most people have some general belief in the objective efficacy of prayer, but none seem willing to admit its action in those special cases of which they have scientific cognisance. Those who may wish to pursue these inquiries upon the effect of prayers for the restoration of health could obtain abundant materials from hospital cases, and in a different way from that proposed in the challenge to which I referred at the beginning of these pages. There are many common maladies whose course is so thoroughly well understood as to admit of accurate tables of probability being constructed for their duration and result. Such are fractures and amputations. Now it would be perfectly practicable to select out of the patients at different hospitals under treatment for fractures and amputations two considerable groups; the one consisting of markedly religious and piously befriended individuals, the other of those who were remarkably cold-hearted and neglected. An honest comparison of their respective periods of treatment and the results would manifest a distinct proof of the efficacy of prayer, if it existed to even a minute fraction of the amount that religious teachers exhort us to believe. An inquiry of a somewhat similar nature may be made into the longevity of persons whose lives are prayed for; and that of the praying classes generally; and in both these cases we can easily obtain statistical facts. The public prayer for the sovereign of every state, Protestant and Catholic, is and has been in the spirit of our own, "Grant her in health long to live." Now, as a simple matter of fact, has this prayer any efficacy? There is a memoir by Dr. Guy, in the *Journal of the Statistical Society* (vol. xxii. p. 355), in which he compares the mean age of sovereigns with that of other classes of persons.[2] His results are expressed in the following table:—

Mean Age attained by Males of various classes who had sur-

[2] Guy 1859.

vived their 30th Year, from 1758 to 1843. Deaths by Accident or Violence are excluded.

			Average	Eminent Men[3]
Members of Royal houses	97	in number	64.04	
Clergy	945		69.49	66.42
Lawyers	294		68.14	66.51
Medical Profession	244		67.31	67.07
English aristocracy	1,179		67.31	
Gentry	1,632		70.22	
Trade and commerce	513		68.74	
Officers in the Royal Navy	366		68.40	
English literature and science	395		67.55	65.22
Officers of the Army	569		67.07	
Fine Arts	239		65.96	64.74

The sovereigns are literally the shortest lived of all who have the advantage of affluence. The prayer has therefore no efficacy, unless the very questionable hypothesis be raised, that the conditions of royal life may naturally be yet more fatal, and that their influence is partly, though incompletely, neutralised by the effects of public prayers.

It will be seen that the same table collates the longevity of clergy, lawyers, and medical men. We are justified in considering the clergy to be a far more prayerful class than either of the other two. It is their profession to pray, and they have the practice of offering morning and evening family prayers in addition to their private devotions. A reference to any of the numerous published collections of family prayers will show that they are full of petitions for temporal benefits. We do not, however, find that the clergy are in any way more long lived in consequence. It is true that the clergy, as a whole, show a life-value of 69.49, as against 68.14 for the lawyers, and 67.31 for the medical men; but the easy country life and

[3] FG: The eminent men are those whose lives are recorded in Chalmers's Biography, with some additions from the Annual Register.

family repose of so many of the clergy are obvious sanatory conditions in their favour. This difference is reversed when the comparison is made between distinguished members of the three classes—that is to say, between persons of sufficient note to have had their lives recorded in a biographical dictionary. When we examine this category, the value of life among the clergy, lawyers, and medical men is as 66.42, 66.51, and 67.04[4] respectively, the clergy being the shortest lived of the three. Hence the prayers of the clergy for protection against the perils and dangers of the night, for protection during the day, and for recovery from sickness, appear to be futile in result.

In my work on "Hereditary Genius" and in the chapter on "Divines," I have worked out the subject with some minuteness on other data, but with precisely the same result. I show that the divines are not specially favoured in those worldly matters for which they naturally pray, but rather the contrary, a fact which I ascribe in part to their having, as a class, indifferent constitutional vigour. I give abundant reason for all this, and do not care to repeat myself; but I should be glad if such of the readers of this present paper who may be accustomed to statistics would refer to the chapter I have mentioned. They will find it of use in confirming what I say here. They will believe me the more when I say that I have taken considerable pains to get at the truth in the questions raised in this present memoir, and that, when I was engaged upon them, I worked, so far as my material went, with as much care as I gave to that chapter on "Divines;" and lastly, they will understand that, when writing the chapter in question, I had all this material by me unused, which justified me in speaking out as decidedly as I did then. A further inquiry may be made into the duration of life among missionaries. We should lay greater stress upon their mortality than upon that of the clergy, because the laudable object of a missionary's career is rendered almost nugatory by his early death. A man goes, say to a tropical climate, in the prime of manhood, who has the probability of many years of useful life before him, had he remained at home. He has the certainty of being able to accomplish sterling good as a missionary, if he should live long enough to learn the language and habits of the country. In the interval he is almost useless. Yet the painful experience of many years shows only too clearly that the missionary is not

[4] The table has 67.07 for medical men.

supernaturally endowed with health. He does not live longer than other people. One missionary after another dies shortly after his arrival. The work that lay almost within the grasp of each of them lingers incompleted.

It must here be repeated, that comparative immunity from disease compels the suspension of no purely material law, if such an expression be permitted. Tropical fever, for example, is due to many subtle causes which are partly under man's control. A single hour's exposure to sun, or wet, or fatigue, or mental agitation, will determine an attack. Now even if God acted only on the minds of the missionaries, his action might be as much to the advantage of their health as if he wrought a physical miracle. He could disincline them to take those courses which might result in mischance, such as the forced march, the wetting, the abstinence from food, or the night exposure, any one of which was competent to develop fever that struck them down. We must not dwell upon the circumstances of individual cases, and say "this was a providential escape," or "that was a salutary chastisement," but we must take the broad averages of mortality, and, when we do so, we find that the missionaries do not form a favoured class.

The efficacy of prayer may yet further be tested by inquiry into the proportion of deaths at the time of birth among the children of the praying and the non-praying classes. The solicitude of parents is so powerfully directed towards the safety of their expected offspring as to leave no room to doubt that pious parents pray fervently for it, especially as death before baptism is considered a most serious evil by many Christians. However, the distribution of still-births appears wholly unaffected by piety. The proportion, for instance, of the still-births published in the Record newspaper and in the Times was found by me, on an examination of a particular period, to bear an identical relation to the total number of deaths. This inquiry might easily be pursued by those who considered that more ample evidence was required.

When we pray in our Liturgy "that the nobility may be indued with grace, wisdom, and understanding," we pray for that which is clearly incompatible with insanity. Does that frightful scourge spare our nobility? Does it spare very religious people more than others? The answer is an emphatic negative to both of these questions. The nobility, probably from their want of the wholesome restraints felt in humbler walks of life, and from their intermarriages, and the very religious people of all denominations, probably from their meditations on hell, are peculiarly subject to it. Religious madness is very common indeed.

As I have already hinted, I do not propose any special inquiry whether the general laws of physical nature are ever suspended in fulfilment of prayer: whether, for instance, success has attended the occasional prayers in the Liturgy when they have been used for rain, for fair weather, for the stilling of the sea in a storm, or for the abatement of a pestilence. I abstain from doing so for two reasons.

First, if it is proved that God does not answer one large class of prayers at all, it would be of less importance to pursue the inquiry. Secondly, the modem feeling of this country is so opposed to a belief in the occasional suspension of the general laws of nature, that an English reader would merely smile at such an investigation. If we are satisfied that the actions of man are not influenced by prayer, even through the subtle influences of his thoughts and will, the only probable form of agency will have been disproved, and no one would care to advance a claim in favour of direct physical interferences.

Biographies do not show, that devotional influences have clustered in any remarkable degree round the youth of those who, whether by their talents or social position, have left a mark upon our English history. Lord Campbell, in his preface to the "Lives of the Chancellors," says, "There is no office in the history of any nation that has been filled with such a long succession of distinguished and interesting men as the office of Lord Chancellor," and that "generally speaking, the most eminent men, if not the most virtuous, have been selected to adorn it." His implied disparagement of their piety is fully sustained by an examination of their respective biographies, and by a taunt of Horace Walpole, quoted in the same preface. An equal absence of remarkable devotional tendencies may be observed in the lives of the leaders of great political parties. The founders of our great families too often owed their advancement to tricky and time-serving courtiership. The belief so frequently expressed in the Psalms, that the descendants of the righteous shall continue, and that those of the wicked shall surely fail, is not fulfilled in the history of our English peerage. Take for instance the highest class, that of the Ducal houses. The influence of social position in this country is so enormous that the possession of a dukedom is a power that can hardly be understood without some sort of calculation. There are, I believe, only twenty-seven dukes to about eight millions of adult male Englishmen, or about three dukes to each million, yet the cabinet of fourteen ministers which governs this country, and India too, commonly contains one duke, often two, and in recent times three. The political privilege inherited with a dukedom in this country is at the lowest estimate many thousand-fold above the average birthright of Englishmen. What was the origin of these ducal families whose influence on the destiny of England and her dependencies is so enormous? Were their founders the eminently devout children of eminently pious parents? Have they and their ancestors been distinguished among the praying classes? Not so. I give in a footnote[5] a list of their names, which recalls many a deed of patriotism, valour, and skill, many an instance of eminent merit of the worldly sort, which we Englishmen honour six days out of the seven—many scandals, many a disgrace, but not, on the other hand, a single instance known to me of eminently prayerful qualities. Four at least of the existing ducal houses are unable to claim the title of having been raised into existence through the devout habits of their progenitors, because the families of Buccleuch, Grafton, St. Albans, and Richmond were thus highly ennobled solely on the ground of their being descended from Charles II. and four of his mistresses, namely, Lucy Walters, Barbara Villiers, Nell Gwynne, and Louise de Querouaille. The dukedom of Cleveland may almost be reckoned as a fifth instance.

The civil liberty we enjoy in England, and the energy of our race, have given rise to a number of institutions, societies, commercial adventures, political meetings, and combinations of all sorts. Some of these are exclusively clerical, some lay, and others mixed. It is impossible for a person to have taken an active share in social life without having had abundant means of estimating for himself, and of hearing the opinion of others, on the value of a preponderating clerical element in business committees. For my own part, I never heard a favourable one. The procedure of Convocation, which, like all exclusively clerical meetings, is opened with prayer, has not inspired the outer world with much respect. The histories

---

[5] FG: Abercorn, Argyll, Athole, Beaufort, Bedford, Buccleuch, Buckingham, Cleveland, Devonshire, Grafton, Hamilton, Leeds, Leinster, Manchester, Marlborough, Montrose, Newcastle, Norfolk, Northumberland, Portland, Richmond, Roxburghe, Rutland, St Albans, Somerset, Sutherland, Wellington.

of the great councils of the Church are most painful to read. There is reason to expect that devout and superstitious men should be unreasonable; for a person who believes his thoughts to be inspired, necessarily accredits his prejudices with divine authority. He is therefore little accessible to argument, and he is intolerant of those whose opinions differ from his, especially on first principles. Consequently, he is a bad coadjutor in business matters. It is a common week-day opinion of the world that praying people are not practical.

Again, there is a large class of instances where an enterprise on behalf of pious people is executed by the agency of the profane. Do such enterprises prosper beyond the average? For instance, a vessel on a missionary errand is navigated by ordinary seamen. A fleet, followed by the prayers of the English nation, carries reinforcements to quell an Indian mutiny. We do not care to ask whether the result of these prayers is to obtain favourable winds, but simply whether they ensue in a propitious voyage, whatever may have been the agencies by which that result was obtained. The success of voyages might be due to many other agencies than the suspension of the physical laws that control the winds and currents; just as we showed that a rapid recovery from illness might be due to other causes than direct interference with cosmic order. It might have been put into the captain's heart to navigate in that course and to perform those acts of seamanship which proved links in a chain that led to eventual success. A very small matter would suffice to make a great difference in the end. A vessel navigated by a man who was a good forecaster of weather and an accomplished hydrographer would considerably outstrip another that was deficient in so accomplished a commander, but otherwise similarly equipped. The perfectly instructed navigator would deviate from the most direct course by perhaps some mere trifle, first here, then there, in order to bring his vessel within favouring slants of wind and advantageous currents. A ship commanded by a captain and steered by a sailor whose hearts were miraculously acted upon in answer to prayer would unconsciously, as by instinct, or even as it were by mistake, perform these deviations from routine, which would lead to ultimate success.

The missionaries who are the most earnestly prayed for are usually those who sail on routes where there is little traffic, and therefore where there is more opportunity for the effects of secret providential overruling to display themselves than among those who sail in ordinary sea voyages. In the usual sea routes a great deal is known of the peculiarities of the seasons and currents, and of the whereabouts of hidden dangers of all kinds; their average risk is small, and the insurance is low. But when vessels are bound to ports like those, sought by the missionaries the case is different. The risk that attends their voyages is largely increased, and the insurance is proportionately raised. But is the risk equally increased in respect to missionary vessels and to those of traders and of slave-dealers? The comparison between the fortune that attends prayerful and non-prayerful people may here be most happily made. The missionaries are eminently among the former category, and the slave-dealers and the traders we speak of in the other. Traders in the unhealthy and barbarous regions to which we refer are notoriously the most godless and reckless (on the broad average) of any of their set. We have, unfortunately, little knowledge of the sea risks of slavers, because the rates of their insurance involve the risk of capture. There is, however, a universal testimony, in the parliamentary reports on slavery, to the excellent and skilful manner in which these vessels are sailed and navigated, which is a *prima facie* reason for believing their sea risks to be small. As to the relative risks run by ordinary traders and missionary vessels, the insurance offices absolutely ignore the slightest difference between them. They look to the class of the vessel, and to the station to which she is bound, and to nothing else. The notion that a missionary or other pious enterprise carries any immunity from danger has never been entertained by insurance companies.

To proceed with our inquiry, whether enterprises on behalf of pious people succeed better than others when they are intrusted to profane hands, we may ask,—Is a bank or other commercial undertaking more secure when devout men are among its shareholders,—or when the funds of pious people, or charities, or of religious bodies are deposited in its keeping, or when its proceedings are opened with prayer, as was the case with the disastrous Royal British Bank? It is impossible to say yes. There are far too many sad experiences of the contrary.

If prayerful habits had influence on temporal success, it is very probable, as we must again repeat, that insurance offices, of at least some descriptions, would long ago have discovered and made allowance for it. It would be most unwise, from a business point of view, to allow the devout, supposing their greater longevity even probable, to obtain annuities at the same low rates as the profane. Before insurance offices accept a life, they make confidential inquiries into the antecedents of the applicant. But such a question has never been heard of as, "Does he habitually use family prayers and private devotions?" Insurance offices, so wakeful to sanatory influences, absolutely ignore prayer as one of them. The same is true for insurances of all descriptions, as those connected with fire, ships, lightning, hail, accidental death, and cattle sickness. How is it possible to explain why Quakers, who are most devout and most shrewd men of business, have ignored these considerations, except on the ground that they do not really believe in what they and others freely assert about the efficacy of prayer? It was at one time considered an act of mistrust in an overruling Providence to put lightning conductors on churches; for it was said that God would surely take care of his own. But Arago's collection of the accidents from lightning showed they were sorely needed; and now lightning conductors are universal. Other kinds of accidents befall churches, equally with other buildings of the same class; such as architectural flaws, resulting in great expenses for repair, fires, earthquakes, and avalanches.

The cogency of all these arguments is materially increased by the recollection that many items of ancient faith have been successively abandoned by the Christian world to the domain of recognised superstition. It is not two centuries ago, long subsequent to the days of Shakespeare and other great names, that the sovereign of this country was accustomed to lay hands on the sick for their recovery, under the sanction of a regular Church service, which was not omitted from our prayer-books till the time of George II. Witches were unanimously believed in, and were regularly exorcised, and punished by law, up to the beginning of the last century. Ordeals and duels, most reasonable solutions of complicated difficulties according to the popular theory of religion, were found absolutely fallacious in practice. The miraculous power of relics and images, still so general in Southern Europe, is scouted in England. The importance ascribed to dreams, the barely extinct claims of astrology, and auguries of good or evil luck, and many other well-known products of superstition which are found to exist in every country, have ceased to be believed in by us. This is the natural course of events, just as the Waters of Jealousy and the

Urim and Thummin of the Mosaic law had become obsolete in the times of the later Jewish kings. The civilised world has already yielded an enormous amount of honest conviction to the inexorable requirements of solid fact; and it seems to me clear that all belief in the efficacy of prayer, in the sense in which I have been considering it, must be yielded also. The evidence I have been able to collect bears wholely[6] and solely in that direction, and in the face of it the *onus probandi* lies henceforth on the other side.

Nothing that I have said negatives the fact that the mind may be relieved by the utterance of prayer. The impulse to pour out the feelings in sound is not peculiar to man. Any mother that has lost her young, and wanders about moaning and looking piteously for sympathy, possesses much of that which prompts men to pray in articulate words. There is a yearning of the heart, a craving for help, it knows not where, certainly from no source that it sees. Of a similar kind is the bitter cry of the hare, when the greyhound is almost upon her; she abandons hope through her own efforts, and screams,—but to whom? It is a voice convulsively sent out into space, whose utterance is a physical relief. These feelings of distress and of terror are simple, and an inarticulate cry suffices to give vent to them; but the reason why man is not satisfied by uttering inarticulate cries (though sometimes they are felt to be the most appropriate) is owing to his superior intellectual powers. His memory travels back through interlacing paths, and dwells on various connected incidents; his emotions are complex, and he prays at length.

Neither does anything I have said profess to throw light on the question of how far it is possible for man to commune in his heart with God. We know that many persons of high intellectual gifts and critical minds look upon it as an axiomatic certainty that they possess this power, although it is impossible for them to establish any satisfactory criterion to distinguish between what may really be borne in upon them from without and what arises from within, but which, through a sham of the imagination, appears to be external. A confident sense of communion with God must necessarily rejoice and strengthen the heart, and divert it from petty cares; and it is equally certain that similar benefits are not excluded from those who on conscientious grounds are sceptical as to the reality of a power of communion. These can dwell on the

---

[6] A rare variant of 'wholly'.

undoubted fact, that there exists a solidarity between themselves and what surrounds them, through the endless reactions of physical laws, among which the hereditary influences are to be included. They know that they are descended from an endless past, that they have a brotherhood with all that is, and have each his own share of responsibility in the parentage of an endless future. The effort to familiarise the imagination with this great idea has much in common with the effort of communing with a God, and its reaction on the mind of the thinker is in many important respects the same. It may not equally rejoice the heart, but it is quite as powerful in ennobling the resolves, and it is found to give serenity during the trials of life and in the shadow of approaching death.

Francis Galton.

Chapter 8

Letter to the *Spectator* on the Efficacy of Prayer[1]

Sir,—I thank you for having opened your columns to discussion upon the efficacy of Prayer, and to have so well acted as moderator in a matter which deeply touches the feelings of many men as to have enabled the discussion to be carried on with mutual forbearance and respect on the part of the disputants.

My object in writing now is to endeavour to confine the discussion to what I conceive to be strategic points, though they are commonly neglected, and are usually, but indirectly, aimed at by your numerous correspondents. Those who deny the right of appeal to statistical inquiries upon the efficacy of prayer assume implicitly two propositions, both of which I gainsay, and which I will now explicitly state. They assert, first, that the desire to pray is intuitive to man (let the word pass, for the moment); secondly, that the cogency of intuition is greater than of observation. I maintain, on the other hand, that the desire to pray is not intuitive, and even if it were, that the cogency of intuition is less than that of observation. As regards the meaning I assign in this letter to "intuitive perceptions," I am perfectly willing to accept the widest definition my adversaries can reasonably desire. I do not wish to haggle about narrowing the limit; it is in no way necessary to my argument that I should do so, therefore I will concede enormously, and will allow that all perceptions or feelings strongly developed in the average man may be reckoned as intuitive to the human race. Now I assert that the desire for prayer is not one of these feelings, but that it is an artificial creation

[1] Galton 1872d.

of theologians; also that the class of similar feelings which are intuitive are such as obedience to dreams, incantations, and witchcraft, fear of the evil eye, belief in demoniacal possession, exorcising, coercion of an angry spirit by some tom-tom ceremony, fetish-worship, and tabu. The savage does not pray by natural inclination, but the missionary teaches him to pray, and as, at the same time, he preaches to him on the existence of a God who listens to prayer, precept to pray is a logical sequence of that instruction. The savage believes in what the missionary tells him, because the missionary is avowedly a more instructed man than himself in many things, and he is certainly in earnest, therefore the missionary's deity is accepted by the savage, and the converted heathen is taught to pray.

In modern civilisation the action of the mother upon the belief and habit of the child resembles in many respects that of the missionary upon those of the savage. She tells him loving tales about God's watchful care, and of His answers to those who kneel and speak to Him, and she joins his little hands together, and sets him on his knees, and teaches him, with caressing earnestness, to pray for temporal blessings, from the very dawn of his intelligence. What wonder that this nursery theology should pervade his life, and that it should be so associated with his deepest feelings that he should at last believe it to have been intuitive? His belief is confirmed by the events of his after life, for on all its solemn occasions it is the habit for the clergyman to step in, and to consecrate them by prayer. He is present by the death-bed, by the marriage altar, and by the baptismal font; he usually superintends early instruction; and he has by custom the opportunity and unrestrained right of preaching and praying before large congregations on every seventh day. Again, I ask, what wonder is it that a habit of prayer and a sense of its necessity should be formed which seem, until their sources have been analysed, to be of primeval origin? My second point is easily disposed of, namely, that even if the belief in prayer were intuitive, its cogency ought to be considered inferior to that which is prompted by the observation of facts. My argument is this,—I do not care to go into the metaphysics of the matter, but would simply point out that the very theologians who insist on the supreme authority of religious intuition are precisely the men who have already most prominently denied it in practice. Their predecessors, at the time of the Christian era, and for hundreds of years subsequently, nay, even men of the present time in Catholic countries, have believed in the divine origin of dreams and auguries, in ordeal and in duel, in lots after prayer, in blessings and in cursings, in witchcraft, in miraculous cures, in demoniacal possessions and in exorcisms. All this the theologians of the present English Church have quietly suppressed, as of "superstitious" origin. They also complacently ignore that their predecessors have been beaten along their whole line by statistical inquiries, for it is by more or less unconscious use of statistics that the belief in ordeal, duel, augury, and the rest, has disappeared, and now that theologians are summoned on statistical grounds to surrender a belief which I have shown to have much less claim to be considered as intuitive, they start with naive indignation, as at a previously unheard-of and most unreasonable interference. You will observe that the views advanced in this letter could be much more strongly enforced by an elaborate essay, but "sapienti verbum sat.," and I write concisely, at the risk of weakening my case, in order to induce those who may answer me in your columns to be equally concise and pointed.—I am, Sir, &c.,

Francis Galton.

Chapter 9

On Blood-Relationship[1]

I propose in this memoir to deduce, by fair reasoning from acknowledged facts, a more definite notion than now exists of the meaning of the word "kinship." It is my aim to analyze and describe the complicated connexion that binds an individual, hereditarily, to his parents and to his brothers and sisters, and, therefore, by an extension of similar links, to his more distant kinsfolk. I hope by these means to set forth the doctrines of heredity in a more orderly and explicit manner than is otherwise practicable.

From the well-known circumstance that an individual may transmit to his descendants ancestral qualities which he does not himself possess, we are assured that they could not have been altogether destroyed in him, but must have maintained their existence in a latent form. Therefore each individual may properly be conceived as consisting of two parts, one of which is latent and only known to us by its effects on his posterity, while the other is patent, and constitutes the person manifest to our senses.

The adjacent and, in a broad sense, separate lines of growth in which the patent and latent elements are situated, diverge from a common group and converge to a common contribution, because they were both evolved out of elements contained in a structureless ovum, and they, jointly, contribute the elements which form the structureless ova of their offspring.

The annexed diagram illustrates my meaning, and serves to show clearly that the span of each of the links in the general chain of heredity extends from one structureless stage to another, and not from person to person:—

[1] Galton 1872a.

Structureless elements in Father	.........Adult Father	Structureless elements in offspring.
	.........Latent in Father.........	

Figure 9.1: Stages of Heredity

I will now proceed to consider the quality of the several relationships by which the above terms are connected together.

The observed facts of Reversion enable us to prove that the latent elements must be greatly more varied than those that are personal or patent. The arguments are as follows:—(1) there must be room for very great variety, because a single strain of impure blood will reassert itself after more than eight generations; (2) an individual has 256 progenitors in the eighth degree, if there have been no ancestral intermarriages, while under the ordinary conditions of social and neighbourly life he will certainly have had a considerable, though a smaller number of them; (3) the gradual waning of the tendency to reversion as the generations increase conforms to what would occur if each fresh marriage contributed a competing element for the same place, thus diluting the impure strain until its relative importance was reduced to an insignificant amount. It follows from these arguments that for each place among the personal elements there may exist, and probably often does exist, a great variety of latent elements that formerly competed to fill it.

I have spoken of the primary elements as they exist in the newly impregnated ovum, where they are structureless but contain the materials out of which structure is evolved; the embryonic elements are segregated from among them. On what principle are they segregated? Since for each place there have been many unsuccessful but qualified competitors, it must have been on some principle whose effects may be described as those of "*Class Representation*," using that phrase in a perfectly general sense as indicating a mere fact, and avoiding any hypothesis or affirmation on points of detail, about most, if not all, of which we are profoundly ignorant. I give as broad a meaning to the expression as a politician would give to the kindred one, a "representative assembly." By this he means to say that the assembly consists of representatives from various constituencies, which is a distinct piece of information so far as it goes, and is a useful one, although it deals with no matter of detail; it says nothing about the number of electors, their qualifications, or the motives by which they are influenced; it gives no information as to the number of seats; it does not tell us how many candidates there are usually for each seat, nor whether the same person is eligible for, or may represent at the same time, more than one place, nor whether the result of the elections at one place may or may not influence those at another (on the principle of correlation). After these explanations there can, I trust, be no difficulty in accepting my definition of the general character of the relation between the embryonic and the structureless elements, that the former are the result of election from the latter on some method of Class Representation.

The embryonic elements are *developed* into the adult person. "Development" is a word whose meaning is quite as distinct in respect to form, and as vague in respect to detail, as the phrase we have just been considering; it embraces the combined effects of growth and multiplication, as well as those of modification in quality and proportion, under both internal and external influences. If we were able to obtain an approximate knowledge of the original elements, statistical experiences would no doubt enable us to predict the average value of the form into which they would become developed, just as a knowledge of the seeds that were sown would enable us to predict in a general way the appearance of the garden when the plants had grown up; but the individual variation of each case would of course be great, owing to the large number of variable influences concerned in the process of development.

The latent elements in the embryonic stage must be developed by a parallel, I do not say by an identical process, into those of the adult stage. Therefore, to avoid all chance of being misapprehended when I collate them, I will call, in the diagram I am about to give (see fig. 1), the one process "Development a" and the other "Development b".

It is not intended to affirm, in making these subdivisions, that the embryonic and adult stages are distinctly separated; they are continuous, and it is impossible but that they should overlap, some elements remaining embryonic while others are completely formed. Nevertheless the two, speaking broadly, may fairly be looked upon as consecutive.

Again, the two processes are not wholly distinct; on the contrary, the embryo, and even the adult in some degree, must receive supplementary contributions derived from their contemporary latent elements, because ancestral qualities indicated in early life frequently disappear and yield place to others. The reverse process is doubtful; it may exist in the embryonic stage, but it certainly does not exist in a sensible degree in the adult stage, else the later children of a union would resemble their parents more nearly than the earlier ones.

Lastly, I must guard myself against the objection that though structure is largely correlated, I have treated it too much as consisting of separate elements. To this I answer, first, that in describing how the embryonic are derived from the structureless elements, I expressly left room for a small degree of correlation; secondly, that in the development of the adult elements from the embryonic there is a perfectly open field for natural selection, which is the agency by which correlation is mainly established; and, thirdly, that correlation affects groups of elements rather than the complete person, as is proved by the frequent occurrence of small groups of persistent peculiarities, which do not affect the rest of the organism, so far as we know, in any way whatever.

The ground we have already gained may be described as follows:—

Out of the structureless ovum the embryonic elements are taken by *Class Representation*, and these are *developed* (a) into the visible adult individual; on the other hand, returning to our starting-point at the structureless ovum, we find, after the embryonic elements have been segregated, the large *Residue* is *developed* (b) into the latent elements contained in the adult individual. All this is summarily expressed in the first two columns of the diagram (fig. 1). I might have inserted vertical arrows to show the minor connexions between the corresponding stages in the two parallel processes, but it would have complicated the figure.

In what way do the patent and latent adult elements respectively contribute representatives towards the structureless stage of the next generation? We know that every quality they possess *may* be transmitted to it, but it does not follow that they *are* invariably transmitted. The contributions from the patent elements cannot be by "Class," because their own original elements have been themselves specialized, and therefore can contain no more than one or a few members of each class (which, it is true, must have been somewhat developed, both in numbers and variety, into what we may call "families"). Their contributions may therefore be justly described as being effected on some principle that has resulted in a "*Family representation*," though whether in the representation of every family I do not profess to say.

As regards the large variety of adult latent elements, they cannot all be transmitted, for the following obvious reason—the corresponding qualities of no two parents can be considered exactly alike; therefore the accumulation of subvarieties, if they were all preserved as the generations rolled onwards, would exceed in multitude the wildest flights of rational theory. The heritage of peculiarities through the contributions of 1000 consecutive generations, even supposing a great deal of ancestral intermarriage, must far exceed what could be packed into a single ovum. The contributions from the latent adult elements are therefore no more than *Representative*; but they have to furnish all the various members of each Class whence its representatives have afterwards to be drawn. Therefore, bearing in mind what has been just argued, that it is impossible for the elements of every individual quality to be contributed, we are driven to suppose, as in the previous case, a "*Family representation*," the similar elements contributed by the two parents ranking, of course, as of the same family. It is most important to bear in mind that this phrase states a fact and not an hypothesis; it does not mean that each and every Family has just one representative, for it is absolutely reticent on all such matters of detail as those I enumerated when speaking of Class Representation. To show the importance which I attach to this disclaimer, I may be permitted to mention what appears to me the most probable *modus operandi*, namely, that it is in reality a large selection made out of larger and not out of smaller constituencies than those I have called "classes," similar to that which would be obtained by an indiscriminate conscription: thus, if a large army be drawn from the provinces of a country by a general conscription, its constitution, according to the laws of chance, will reflect with surprising precision the qualities of the population whence it was taken; each village will be found to furnish a contingent, and the composition of the army will be sensibly the same as if it had been due to a system of immediate representation from the several villages.

The diagram (fig. 1) expresses the whole of the foregoing results; it begins with the structureless elements whence the parent individual was formed, and ends with his contributions to the structureless elements whence his offspring is formed.

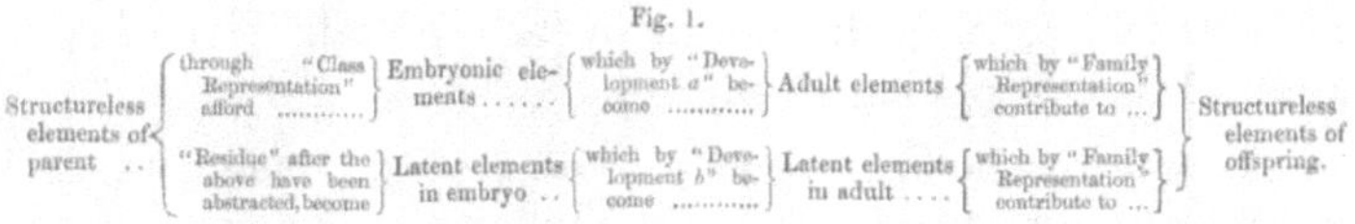

Figure 9.2: Galton's Figure 1

I will now inquire what are, roughly speaking, the relative proportions of the contributions to the elements of the offspring made respectively by the patent and latent elements of the adult parent. It is better not to complicate the inquiry by speaking, at first, of these elements in their entirety, but rather of some special characteristic: thus, to fix the ideas, suppose we are speaking about a peculiar skin-mark in an animal; the peculiarity in question may be conceived (1) as purely personal, without the concurrence of any latent equivalents, (2) as personal but conjoined with latent equivalents, and (3) as existent wholly in a latent form. It can be shown that, in the first case, the power of hereditary transmission is exceedingly feeble; for, notwithstanding some exceptions (as in the lost power of flight in domestic birds), the effects of the use and disuse of limbs, and those of habit, are transmitted to posterity in only a very slight degree. Again, it can be fairly argued that many instances which seem at first sight to fall under case (1), that is, to be purely personal, and to prove a larger hereditary influence than what I assign to it, do really belong to case (2): thus, when individuals born with a peculiar mark are reputed to be the first of their race in whom it had ever appeared, it would be hazardous in the extreme to argue that the latent elements of that mark were wholly deficient in them. It is very remarkable (I was indebted for a knowledge of this fact to Mr. Tegetmeier[2]) how nearly every bar or spot found in any species of an animal in its wild state may be bred into existence in the domesticated variety of that species, showing that the elements of all these bars and spots are universally present in all varieties of the species, though their manifestation may be overborne and suppressed. We therefore see that the hereditary influences of an animal with respect to any particular spot are, I will not say in every case, but certainly on the *average of many cases*, much more numerous than if that

---

[2] The ornithologist William Bernhardt Tegetmeier (1816–1912).

spot had been purely a personal characteristic, without the concurrence of any latent elements. Bearing this argument in mind, we shall more justly estimate the import of the statistical evidence to be obtained from breeders of animals. I should judge, from the impression left by many scattered statistics, that it is perfectly safe to affirm that breeders, when they mate two animals, each having the same unusual characteristic, not through known hereditary transmission, but by supposed variation, would consider themselves fortunate if one quarter of the progeny inherited that quality. Now these successful cases are, as I have shown, on the average, the produce of parents having the peculiarity not only in a personal but also, to some degree, in a latent form. We may therefore reasonably conclude that, had the latter portion been non-existent, the ratio of successful cases would have been materially diminished.

I should demur, on precisely the same grounds, to objections based on the fact of the transmission of qualities to grandchildren being more frequent through children who possess those qualities than through children who do not; for I maintain that the personal manifestation is, on the average, though it need not be so in every case, a certain proof of the existence of some latent elements.

Having proved how small is the power of hereditary transmission of the personal elements, we can easily show how large is the transmission of the purely latent elements, in the case (3), by appealing to the well-known facts of Reversion; but into these it is hardly necessary for me to enter at length. The general and safe conclusion is, that the contribution from the patent elements is very much less than from the latent

If we now combine our results into a diagram (fig. 2), showing the fainter streams of heredity by *italic lines*, and indicating those processes by asterisks (*) which were described at length in the previous figure, we shall easily recognize the complexity of hereditary problems. We see that parents are very indirectly and only partially related to their own children, and that there are two lines of connexion between them, the one of large and the other of small relative importance. The former is a collateral kinship and very distant, the parent being descended through two stages (two asterisks) from a structureless source, and the child (so far as that parent is concerned) through five totally distinct stages from the same source; the other, but unimportant line of connexion, is direct and connects the child with the parent through two stages.

We shall therefore wonder that, notwithstanding the fact of an average resemblance between parent and child, the amount of individual variation should not be much greater than it is, until we have realized how complete must be the harmony between every variety and its environments in order that the variety should be permanent.

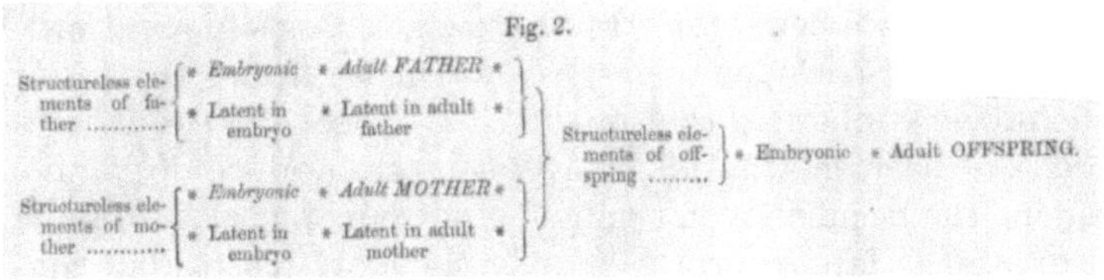

Figure 9.3: Galton's Figure 2

We also infer from the diagram how much nearer, and yet how subject to variation, is the kinship between the children of the same parents; for only two stages are required to trace back their descent to a common origin, which, however, proceeds from four separate streams of heredity, namely the adult patent and latent elements of each of the two parents.

An approximate notion of the nearest conceivable relationship between a parent and his child may be gained by supposing an urn containing a great number of balls, marked in various ways, and a handful to be drawn out of them at random as a sample: this sample would represent the person of a parent. Let us next suppose the sample to be examined, and a few handfuls of new balls to be marked according to the patterns of those found in the sample, and to be thrown along with them back into the urn. Now let the contents of another urn, representing the influences of the other parent, be mixed with those of the first. Lastly, suppose a second sample to be drawn out of the combined contents of the two urns, to represent the offspring. There can be no nearer connexion justly conceived to subsist between the parent and child than between the two samples; on the contrary, my diagram shows the relationship to be in reality much more remote, and consisting of many consecutive stages, and therefore hardly to be expressed by such simple chances. Whenever the balls in the urns are much of the same pattern, the samples will be alike, but not otherwise. The offspring of a mongrel stock necessarily deviate in appearance from each other and from their parents.

We cannot now fail to be impressed with the fallacy of reckoning inheritance in the usual way, from parents to offspring, using those words in their popular sense of visible personalities. The span of the true hereditary link connects, as I have already insisted upon, not the parent with the offspring, but the primary elements of the two, such as they existed in the newly impregnated ova, whence they were respectively developed. No valid excuse can be offered for not attending to this fact, on the ground of our ignorance of the variety and proportionate values of the primary elements: we do not mend matters in the least, but we gratuitously add confusion to our ignorance, by dealing with hereditary facts on the plan of ordinary pedigrees—namely, from the *persons* of the parents to those of their offspring.

It will be observed that, owing to the clearer idea we have now obtained of the meaning of kinship and of the consecutive phases of the chain of life, the various causes of individual variation can be easily and surely sorted into their proper places. I will mention a few of them, merely as examples.

Previous to the segregation of the embryonic elements, if the structureless ones be diverse without any strongly preponderating element, it is impossible to foresee the character of the embryo, just as it is impossible to foresee the character of a handful chosen from an urn containing a mixed assemblage of variously coloured balls; but if they be not diverse, then the embryonic elements will be a true sample of the structureless ones, the conditions of purity of blood are fulfilled, and the offspring will resemble its parents.

We also see, in the process by which the embryonic elements are obtained, how the curious phenomenon may occur of inheritance occasionally skipping alternate generations. The more that has been removed from the structureless group for the supply of the embryonic (which, as we have seen, is a nearly sterile destination) the less remains for the "residue," too little, it may be, to assert itself by that, the only prolific, line of transmission. In the supposed case it would recuperate itself during the succeeding generation, where the elements in question will have remained wholly latent, owing to their insignificance in the structureless stage of that generation, which would be sufficient to secure any portion of it from selection for the embryonic form.

Again, it is in the process of selection of elements, both latent and patent, from the adult parents for the structureless stage of the next generation, where I suppose the curious and unknown conditions usually to occur through which a change in the habits of life, after the adult age has been reached, is apt to produce sterility. I may be permitted to remark, hypothetically, that this view appears to be corroborated by the fact that many grains of pollen or many spermatozoa are required to fertilize each ovum, because, as it would seem, each separate one does not contain a sufficiently complete representation of the primary elements to supply the needs of an individual life, and that it is only by the accumulation of several separate consignments (so to speak) of the representative elements that the necessary variety is ensured. I argue from this that there is a tendency to a large individual variation in the constituents of each grain of pollen, or spermatozoon, and, by analogy, that there is a similar though smaller tendency in each ovum; also that changes in the habits of life may increase this variation to a degree that involves sterility.

Lastly, it is often remarked (1) that the immediate offspring of different races or even varieties resemble their parents equally, but (2) that great diversities appear in the next and in succeeding generations. In which stage does the variability occur? It cannot be in the first (class representation) nor in the second (development), else (1) could not have been true; therefore it must be in the third stage. A white parent necessarily contributes white elements to the structureless stage of his offspring, and a black, black; but it does not in the least follow that the contributions from a true mulatto must be truly mulatto.

One result of this investigation is to show very clearly that large variation in individuals from their parents is not incompatible with the strict doctrine of heredity, but is a consequence of it wherever the breed is impure. I am desirous of applying these considerations to the intellectual and moral gifts of the human race, which is more mongrelized than that of any other domesticated animal. It has been thought by some that the fact of children frequently showing marked individual variation in ability from that of their parents is a proof that intellectual and moral gifts are not strictly transmitted by inheritance. My arguments lead to exactly the opposite result. I show that their great individual variation is a necessity under present conditions; and I maintain that results derived from large averages are all that can be required, and all we could expect to obtain, to prove that intellectual and moral gifts are as strictly matters of inheritance as any purely physical qualities.

Chapter 10

[Livingstone, the RGS and Lieut. Dawson]

To the Editor of the Daily News[1]

SIR,—The number and character of the misconceptions I find in the newspapers of subjects on which I can speak with the authority of certain knowledge are simply amazing; but I will confine nearly all of what I am about to say to the letter signed "P." in your impression of yesterday, Sept. 5. It will not, I trust, be thought by my colleagues on the council of the Royal Geographical Society to be a breach of confidence if I state (1.) that the letter of instructions to Lieut. Dawson was draughted by myself, well-considered by a large council and accepted with trifling modifications. (2.) That I acted as chairman of a small sub-committee to arrange about the *materiel* of the expedition; my colleagues being the late political resident in Zanzibar, General Rigby; that most accomplished African geographer, Mr. Findlay (not one of the "critics" about whom I shall have to speak), and Colonel

---

1 *Daily News*, 1872-09-07, 5-6. This is a response to a letter to the editor by 'P.', the Reform Club, published two days earlier, 1872-09-05, 6. Clements Markham had published a letter to the editor, 1872-09-04, 6, which dismissed a report in the *Daily News* that Livingstone was on bad terms with the RGS. Markham responded on 1872-09-06, 5, to the letter from 'P.' as 'a great deal of nonsense', and on 1872-09-09, 5. to a further letter from 'P.' which had been published the same day as Galton's letter on 1872-09-07, 5. That second letter from 'P.' had disclosed the contents of a private letter from Markham to Stanley, leaked by the latter, which tried to placate the reporter, but 'P.' describes as trying to 'square' him. In fact Markham's letter was a response to one from Stanley on 1872-09-05, see GALTON/1/2/4/12, which dangled the prospect of reconciliation if the RGS did something unspecified for the reporter.

Grant (Speke and Grant) whenever he was able to attend. (3.) The private letter of geographical instructions, a roughly-written reminder of what I thought might prove useful hints on surveying matters, which somehow found its way into your columns, was my writing. (4.) I, myself, wrote and proposed the letter of thanks to Mr. Stanley at the Livingstone Search Committee, which, with slight modifications, was sent to him. (5.) I have served on the Council of the Society for many years, and, for some time, held the post of its secretary.

I thus claim some right to interpret the views of the members of the Council in its recent action, which has ended so abortively, and I unhesitatingly deny the existence of any feeling less than one of admiration and pride in Dr. Livingstone as being the most eminent explorer of the day.

The "Insane Geographer"[2] paragraph curiously enough refers, I believe, to one or other of four gentlemen, three of whom have not served on the Council for very many years, if at all—I mean the late Mr. M'Queen, who was, I grant, a provoking critic; Mr. Cooley, who will not be offended, I hope, if I call him a caustic as well as a most learned critic in historical geography; and Dr. Beke, the well-known Abyssinian traveller, and the frequent correspondent of the *Athenaeum* and the *Times* on such points of African geography as may from time to time come under public notice. Besides these, is Livingstone's old friend Mr. Arrowsmith, to whose conscientious and thorough map work all travellers who have had occasion to apply to him eagerly testify, and who brought Livingstone's first map into a shape far more satisfactory to the traveller himself than was the original drawing. This is alluded to in the remarks that preceded the passage in question. It is notorious that the alteration of Lake Nyassa at a place then unseen by Livingstone was violently objected to by him; but it was afterwards, I believe, fully approved of; and his attachment to Mr. Arrowsmith before leaving England for the last time was such, that he insisted on spending his concluding days as a guest at Arrowsmith's house. I believe, but do not know the particulars, that the alteration of a position fixed by astronomical observations (alluded to in the paragraph) was made in ignorance of all the particulars by Dr. Beke. If I am wrong it really matters little, but I hope I shall

[2] According to 'P.', Stanley and Dawson had reported that Livingstone referred to 'a few insane geographers' in the RGS, whose 'interference and criticism' he resented.

be corrected.[3] Now all these matters refer to a transaction, as I am informed, 17 years old. The passage shows to my mind a disposition to have brooded in solitude over trifles until they became of exaggerated moment, and I ascribed its being written to a temporary irritability, likely to pass away and to be regretted. For what more can it matter if a particular geographer interprets numerical data in a wrong way, than if any other person misinterprets anything else? Publicity and discussion are the very things wanted in geography before a map is finally drawn up. I do not believe that the map of any traveller in wild countries extending over a large area bordering on lands visited by other travellers and mapped by them, has ever stood intact after recalculation of positions at home, and analysis of the evidence on which it was constructed. Travellers are usually as sensitive about their conclusions being criticized as experts of all kinds are, when giving evidence on disputed points. Nay, they are more so, from the habits of their solitary life, which prevent their being accustomed to hear contrary opinions without offence, and their learning by frequent experience how much good arises out of discussion. My belief is, that Livingstone is as sensitive as other travellers in these respects, criticising, as they all do, one another unsparingly, and not sufficiently recognising how much they themselves are justly amenable to it. Let us seek for an example in Livingstone's last despatches. It appears he was a resident for a long time on the shores of Lake Tanganyika, and yet firmly believed its waters to run out to the North, to the Albert N'Yanza, through the Rusizi. Yet Mr. Stanley goes to the end of the Lake with him, and finds that the Rusizi runs into the Lake, not out of it, and that the Tanganyika must find an exit somewhere to the southwards of Ujiji. Here is a conclusion of Dr. Livingstone's obviously at fault, showing that geographical conclusions, as distinguished from positive evidence, are open questions. On the other hand, Livingstone criticises other great travellers on their own ground, without compunction, and I do not blame him for it, only I insist that justice should be even-handed. Then in this last despatch he strongly combats Speke and Grant's drawing of the Victoria N'Yanza, and cuts it up into three or four small lakes. But I have little doubt that he would make strong objection if Colonel Grant happened, on the grounds of some native ev-

[3] Beke wrote to the paper that day to deny that he made any alterations, *Daily News*, 1872-09-09.

idence, to make alterations of the same amount in his own theoretical geography. In short, I look upon Livingstone, like most men, to be very ready to criticise the theories of others, and not much to relish criticisms of his own theories; and I looked on the "insane geographer" paragraph as a morbid and exaggerated outburst of this feeling, having a touch of the effects of African fever in it, likely to be regretted hereafter, and not to be seriously regarded by us in the meantime. I also think that if Dr. Livingstone had really intended, before leaving England, to send nothing to the Geographical Society, he should and would have said so before he took their 500*l*. Again, there was much in the prolonged silence of Dr. Livingstone which could not, and cannot yet, so far as I know, be satisfactorily accounted for. My own view was chiefly this—that Livingstone, as I fully believe, is a true hero and a great man; but heroes and great men are never perfect, and he has the peculiarity of entertaining dislikes as violent as they appear to me capricious and wrong-headed. What I have said about Mr. Arrowsmith will serve as one example; and his recent denunciation of his old friend Dr. Kirk, which, to the best of my present knowledge, derived from much independent testimony, is cruelly unjust, will serve as another. I therefore thought it quite possible that Dr. Livingstone might have taken great offence at something. Anyhow I felt sure it would be proper that he should be approached with tact, but that it ought to be unmistakeably pointed out that, from some cause or other, whatever miscarriage of letters or what else, his letters had contained less geography than has hitherto been invariably furnished by travellers circumstanced like himself.

Your correspondent objects that the Royal Geographical Society thought much of geography, and not of the mere relief of Dr. Livingstone. Of course they did. The society is not a humane society, established to succour persons in distress, but for the promotion of geographical science. The 500*l*. voted for the expedition by the Council of the Royal Geographical Society would have been a malappropriation of the society's funds if there was no intention to utilise the expedition, as could perfectly be done, to obtain geographical facts. The public may be considered to have sent their subscriptions mainly, but not entirely, for the relief of a man personally dear to the nation, and the two objects of relief and geography admitted of being simultaneously fulfilled. It is a strange ground of blame of the council of a geographical society that it should show itself zealous to accomplish geographical work.

Now, let there be no misunderstanding as to the chief geographical facts which I, for my part, looked to Lieutenant Dawson to obtain.[4] Those to be derived from Dr. Livingstone would be the observations of latitude, longitude, and altitude above sea level of those few cardinal points which he had already fixed and written about, or was about to describe; also the quantity of water discharged by his rivers. These, if sent, would, according to the custom of the Society, be sent to Greenwich Observatory, to be carefully re-calculated. The originals would be preserved in the Society's rooms, and the public would ever after have virtual access to them. I did not expect that Livingstone would care or like to send back any detail which he hoped shortly to be able to re-touch, but I thought he might well be persuaded to send sealed copies of the original but imperfect observations home, in case of accident. This Livingstone may have done, but there is nothing to show that the "Letts' Diary" contains other than mere narrative.[5] Again, I hoped that Lieutenant Dawson, as a skilful, I might say, professional nautical surveyor, would be able to afford great assistance in overhauling and finding the errors of Livingstone's necessarily battered or otherwise untrustworthy instruments. I also expected he would "pick up the ends" of Livingstone's route at the point where he met him, by rigorously determining its position and elevation; also that he would map wherever he went. I cannot for the moment imagine any reasonable traveller being otherwise than gratified by such appropriate assistance. No doubt, as I have already said, Livingstone ought to be approached with tact; but when the matter had been distinctly explained there would be, to the best of my judgment, absolutely no cause for distrust. It is abundantly certain that these dry details would have no interest to the public; while the graphic letters he has already sent do in some degree forestall his book. And why do we want these dry details? Clearly because Livingstone's researches have a bearing on the routes of adjacent travellers. Exploration is always going on, and it is for us at home to help, by diffusing information useful to explorers. The geographical position of Cazembe's town, supposed to have been fixed by Lacerda, but unfixed 30 miles—we know not in what

---

4 'P.' had complained of the 'absurdly cumbrous instructions' given to Dawson, a direct reference to Galton.

5 Sent by Livingstone to his daughter in England, containing his notes.

direction—by Livingstone, is itself one of those cardinal points by which the routes of numerous traders, who took notes but no observations, can be laid down. Those who argue that Livingstone would be justified in withholding this information after the case was distinctly explained to him, seem to me to say, that he ought to be actuated by the pettiest of personal motives. That which has made Livingstone dear to the nation is his heroic abnegation of self in persistently and laboriously pursuing geographical and anti-slave-trade ends; but by the former more than by the latter. Does it not strike "P." as incongruous that such a man should withhold the very thing for the achievement of which he is so highly honoured? Let me add, that to the best of my knowledge, there has ever been a complete unanimity of feeling among the members of the Council of the Geographical Society, in respect to what was due to Dr. Livingstone.

I have now only to refer to what your correspondent seems to look upon as a want of cordiality on the part of the Council of the Geographical Society in hailing Mr. Stanley as the reliever of Dr. Livingstone. It is unnecessary to recall the circumstances under which the news first arrived, and continued flowing in telegraphic driblets, and the many conflicting opinions afloat. For my part, I thought that until Livingstone's own handwriting and acknowledgment of Mr. Stanley's services reached the country, it would be judicious to take no action. These letters were delivered to various persons on a Thursday and a Friday. Monday is the usual day on which the Geographical committees meet, but the next Monday was the Bank holiday, and it happened that a Livingstone Search Committee had been summoned for the Tuesday. It was on that day and on that occasion that the letter of thanks to Mr. Stanley was decided on and sent, to which I referred in the beginning of this letter, and it was the very earliest opportunity at which it was feasible to send it. The period of the year was one when the Society was in recess, and an adequate meeting of the Council could not be obtained without great inconvenience; but I maintain that the recognition then sent was quite early enough and cordial enough. Livingstone is a consul; but Lord Granville delayed his acknowledgments for finding and relieving his own missing public servant to a much later date. I am confident that the more the public know of all these matters on which the Council of the Society, or its committees, has taken action, the more will they have reason to acknowledge that they have throughout acted with the fairest and most generous motives, that they have taken a vast amount of labour, and that each step was the most promising one at the time when it was taken.—I am, & c.,

Sept. 6
Francis Galton

Chapter 11

Hereditary Improvement.

It is freely allowed by most authorities on heredity,[1] that men are just as subject to its laws, both in body and mind, as are any other animals, but it is almost universally doubted, if not denied, that an establishment of this fact could ever be of large practical benefit to humanity. It is objected that, philosophise as you will, men and women will continue to marry as they have hitherto done, according to their personal likings; that any prospect of improving the race of man is absurd and chimerical, and that though enquiries into the laws of human heredity may be pursued for the satisfaction of a curious disposition, they can be of no real importance. In opposition to these objections, I maintain, in the present essay, that it is feasible to improve the race of man by a system which shall be perfectly in accordance with the moral sense of the present time. I shall first describe the condition, such as I believe it to be, of the existing race of man, and will afterwards propose a scheme for its improvement whose seeds would be planted almost without knowing it, and would slowly but steadily grow, until it had transformed the nation. If the ordinary doctrines of heredity in a broad sense be true, the scheme in question must, as it appears to me, begin to show vigorous life so soon as the mass of educated men shall have learnt to appreciate their truth. But if the doctrines be false, then all I build upon them is of course fallacious.

The bodily and mental condition of every man are, in part, the result of his own voluntary and bygone acts; but experience teaches us that they are also shaped by two other agencies, for neither of which he is responsible; the one, the

[1] Galton 1873b

constitutional peculiarities transmitted to him by inheritance, and the other, the various circumstances to which he has been perforce subjected, especially in early life. Now, in this essay I do not propose to allude to ordinary education, family and national tradition, and other similar moral agencies of high importance. I leave them for the present, to one side; the residue with which alone I am about to deal, may be concisely and sufficiently expressed by the words 'race' and 'nurture.' It is to the consideration of the first of these that the following pages are chiefly devoted; but not entirely so, for I acknowledge that we cannot wholly disentangle their several effects. An improvement in the nurture of a race will eradicate inherited disease; consequently, it is beyond dispute that if our future population were reared under more favourable conditions than at present, both their health and that of their descendants would be greatly improved. There is nothing in what I am about to say that shall underrate the sterling value of nurture, including all kinds of sanitary improvements; nay, I wish to claim them as powerful auxiliaries to my cause; nevertheless, I look upon race as far more important than nurture. Race has a double effect, it creates better and more intelligent individuals, and these become more competent than their predecessors to make laws and customs, whose effects shall favourably react on their own health and on the nurture of their children. The merits and demerits of different races is strongly marked in colonies, where men begin a new life, to a great degree detached from the influences under which they had been reared. Now we may watch a band of Englishmen subjected to no regular authority, but attracted to some new gold-digging, and we shall see that law and order will be gradually evolved, and that the community will purify itself and become respectable, and this is true of hardly any other race of men. Constitutional stamina, strength, intelligence, and moral qualities cling to a breed, say of dogs, notwithstanding many generations of careless nurture; while careful nurture, unaided by selection, can do little more to an inferior breed than eradicate disease and make it good of its kind. Those who would assign more importance to nurture than I have done, must concede that the sanitary conditions under which the mass of the population will hereafter live, are never likely to be so favourable to health as those which are now enjoyed by our wealthy classes. The latter may make many mistakes in matters of health; but they have enormous residual advantages. They can command good food, spacious rooms, and change of air, which is more than equivalent to what the future achievements of sanitary science are likely to afford to the mass of the population. Yet how far are our wealthier classes from the secure possession of those high physical and mental qualities which are the birthright of a good race. Whoever has spent a winter at the health-resorts of the South of France, must have been appalled at witnessing the number of their fellow-countrymen who are afflicted with wretched constitutions, while that of the sickly children, narrow-chested men, and fragile, delicate women who remain at home, is utterly disproportionate to the sickly and misshapen contingent of the stock of any of our breeds of domestic animals.

I need not speak in detail of the many ways in which the forms of civilisation, which have hitherto prevailed, tend to spoil a race, because they must, by this time, have become familiar to all who are interested in heredity; it is sufficient just to allude to two of the chief among those which are now in activity. The first is, the free power of bequeathing wealth, which interferes with the salutary action of natural selection, by preserving the wealthy, and by encouraging marriage on grounds quite independent of personal qualities; and the second is the centralising tendency of our civilisation, which attracts the abler men to towns, where the discouragement to marry is great, and where marriage is comparatively unproductive of descendants who reach adult life. In a paper just communicated to the Statistical Society, I have carefully analysed and discussed the census returns of 1,000 families of factory operatives in Coventry, and of the same number of agricultural labourers in the neighbouring small rural parishes of Warwickshire, and find that the former have little more than half as many adult grandchildren as the latter. They have fewer offspring, and of those few a smaller proportion reach adult life, while the two classes marry with about equal frequency and at about the same ages. The allurements and exigencies of a centralised civilisation are therefore seriously prejudicial to the better class of the human stock, which is first attracted to the towns, and there destroyed; and a system of selection is created whose action is exactly adverse to the good of a race. Again, the ordinary struggle for existence under the bad sanitary conditions of our towns, seems to me to spoil, and not to improve our breed. It selects those who are able to withstand zymotic diseases and impure and insufficient food, but such are not necessarily foremost in the qualities which make a nation great. On the contrary, it is the classes of a coarser organisation who seem to be, on the whole, most favoured under this principle of selection, and who survive to become the parents of the next generation. Visitors to Ireland after the potato famine generally remarked that the Irish type of face seemed to have become more prognathous, that is, more like the negro in the protrusion of the lower jaw; the interpretation of which was, that the men who survived the starvation and other deadly accidents of that horrible time, were more generally of a low and coarse organisation. So again, in every malarious country, the traveller is pained by the sight of the miserable individuals who inhabit it. These have the pre-eminent gift of being able to survive fever, and therefore, by the law of economy of structure, are apt to be deficient in every quality less useful to the exceptional circumstances of their life. The reports of the health of our factory towns disclose a terrible proportion of bad constitutions and invalidism among the operatives, as shown by intermitting pulse, curved spine, narrow chests, and other measurable effects; and at the same time we learn from the census that our population is steadily becoming more urban. Twenty years ago the rural element preponderated; ten years ago the urban became equal to it; and now the urban is in the majority. We have therefore much reason to bestir ourselves to resist the serious deterioration which threatens our race.

I have hitherto addressed myself to the purely physical qualities of mankind, on the importance of which it would have been difficult to have sufficiently insisted a few years ago, when there was a prevailing feeling that the mind was everything and the body nothing. But a reaction has set in, and it has become pretty generally recognised that unless the body be in sound order, we are not likely to get much healthy work or instinct out of it. A powerful brain is an excellent thing, but it requires for its proper maintenance a good pair of lungs, a vigorous heart, and especially a strong stomach, otherwise its outcome of thought is likely to be morbid. This being understood, I will proceed to the mental qualities of our race.

I have written much in my work on *Hereditary Genius* about the average intellect of modern civilised races being unequal to cope with the requirements of the mode of life which circumstances have latterly imposed upon them, and much more might be said on the same subject. The advance in means of communication has made large nations or federations a necessity, whose existence implies a vast number of complicated interests and nice adjustments, which require to be treated in a very intelligent manner, or will otherwise have to be brutally ordered by despotic power. We have latterly seen that the best statesmen of our day are little capable of expressing their meaning in intelligible language, so that political relations are apt to become embroiled by mere misunderstanding of what is intended to be conveyed. In no walk of civilised life do the intellects of men seem equal to what is required of them. It is true that Anglo-Saxons are quite competent to grapple with the everyday problems of small communities, but they have insufficient ability for the due performance of the more difficult duties of citizens of large nations. Consequently, the functions of men engaged in trades and professions of all kinds are adjusted to a dangerously low standard, and the political insight of the multitude goes little deeper than the surface, and is applied in few directions except those to which their guides have pointed. Great nations, instead of being highly organised bodies, are little more than aggregations of men severally intent on self-advancement, who must be cemented into a mass by blind feelings of gregariousness and reverence to mere rank, mere authority, and mere tradition, or they will assuredly fall asunder.

As regards the moral qualities, which are closely interwoven with the intellectual, we cannot but observe the considerable effect which the influence of many generations of civilised life has already exercised upon the race of man. It has already bred out of us many of the wild instincts of our savage forefathers, and has given us a stricter conscience and a larger power of self-control than, judging from the analogy of modern savages, they appear to have had. The possibility of eradicating instinctive wildness, and of introducing an instinctively affectionate disposition into any breed of animals, is clearly proved by what has been effected in dogs. The currish and wolfish nature of such as may be seen roaming at large in the streets of Eastern towns, has been largely suppressed in that of their tamed descendants, who, after many generations of selection and friendly treatment, have also acquired the curious innate love of man to which Mr. Darwin drew attention. All this gives hope for the future of our race, especially if 'viriculture' be possible, notwithstanding that our present moral nature is as unfitted for a high-toned civilisation as our intellectual nature is unfitted to deal with a complex one. It is curious to observe the great variety in the morals of the human race, such as have been delineated by Theophrastus, La Bruyère, and the phrenologists. It seems to me that natural selection has had no influence in securing dominance to the noblest of them, because in the various tactics of the individual battle for life, any one of these qualities in excess may be serviceable to its possessor. But the case would be very different in those higher forms of civilisation, vainly tried as yet, of which the notion of personal property is not the foundation, but which are, in honest truth, republican and co-operative, the good of the community being literally a more vivid desire than, that of self-aggrandisement or any other motive whatever. This is a stage which the human race is undoubtedly destined sooner or later to reach, but which the deficient moral gifts of existing races render them incapable of attaining. It is the obvious course of intelligent men—and I venture to say it should be their religious duty—to advance in the direction whither Nature is determined they shall go; that is, towards the improvement of their race. Thither she will assuredly goad them with a ruthless arm if they hang back, and it is of no avail to kick against the pricks. We are exceedingly blind to the ultimate purposes for which we have come into life, and we know that no small part of the intentions by which we are most apt to be guided, are mere illusions. If, however, we look around at the course of nature, one authoritative fact becomes distinctly prominent, let us make of it what we may. It is, that the life of the individual is treated as of absolutely no importance, while the race is treated as everything, Nature being wholly careless of the former except as a contributor to the maintenance and evolution of the latter. Myriads of inchoate lives are produced in what, to our best judgment, seems a wasteful and reckless manner, in order that a few selected specimens may survive, and be the parents of the next generation. It is as though individual lives were of no more consideration than are the senseless chips which fall from the chisel of the artist who is elaborating some ideal form out of a rude block. We are naturally apt to think of ourselves and of those around us that, being not senseless chips, but living and suffering beings, we should be of primary importance, whereas it seems perfectly clear that our individual lives are little more than agents towards attaining some great and common end of evolution. We must loyally accept the facts as they are, and solace ourselves with such hypotheses as may seem most credible to us. For my part, I cling to the idea of a conscious solidarity in nature, and of its laborious advance under many restrictions, the Whole being conscious of us temporarily detached individuals, but we being very imperfectly and darkly conscious of the Whole. Be this as it may, it becomes our bounden duty to conform our steps to the paths which we recognise to be defined, as those in which sooner or later we have to go. We must, therefore, try to render our individual aims subordinate to those which lead to the improvement of the race. The enthusiasm of humanity, strange as the doctrine may sound, has to be directed primarily to the future of our race, and only secondarily to the well-being of our contemporaries. The ants who, when their nest is disturbed, hurry away each with an uninteresting looking egg, picked up at hazard, not even its own, but not the less precious to it, have their instincts curiously in accordance with the real requirements of Nature. So far as we can interpret her, we read in the clearest letters that our desire for the improvement of our race ought to rise to the force of a passion; and if others interpret Nature in the same way, we may expect that at some future time, perhaps not very remote, it may come to be looked upon as one of the chief religious obligations. It is no absurdity to expect, that it may hereafter be preached, that while helpfulness to the weak, and sympathy with the suffering, is the natural form of outpouring of a merciful and kindly heart, yet that the highest action of all is to provide a vigorous, national life, and that one practical and effective way in which individuals of feeble constitution can show mercy to their kind is by celibacy, lest they should bring beings into existence whose race is predoomed to destruction by the laws of nature. It may come to be avowed as a paramount duty, to anticipate the slow and stubborn processes of natural selection, by endeavouring to breed out feeble constitutions, and petty and ignoble instincts, and to breed in those which are vigorous and noble and social.

The precise problem I have in view, is not only the restoration of the average worth of our race, debased as it has been from its 'typical level' by those deleterious influences of modern civilisation to which I have referred, but to raise it higher still. It has been depressed by those mischievous influences of artificial selection which I have named, and by many others besides. Cannot we, I ask—and I will try to answer the question in the affirmative—introduce other influences which shall counteract and overbear the former, and elevate the race above its typical level at least as much as the former had depressed it? I mean by the phrase 'typical' level the average standard of the race, such as it would become in two or three generations if left unpruned by artificial selection, and if reared under what might be accepted as fair conditions of nurture and a moderate amount of healthy, natural selection. It is to be recollected that individuals are not the offspring of their parents alone, but also of their ancestry to very remote degrees, and that although by a faulty system of civilisation the average worth of a race may become depressed, it has nevertheless an inherent ancestral power of partly recovering from that depression, if a chance be given it of doing so. It has, on the one hand, the advantage of the civilised habits ingrained into its nature, and, on the other hand, it may rise above the abnormal state of depression to which the evil influences of the artificial selection of our modern civilisation have temporarily reduced it.

In my work on *Hereditary Genius* I entered at considerable length upon the classification of men in different grades of natural ability, separated by equal intervals, and showed how we might estimate the proportionate numbers of men in each of them, by availing ourselves of a law, whose traces are to be met with in all the variable phenomena of nature. For example, it will be found that we may divide any body of individuals into four equal groups, of which two shall consist of mediocrities, and the other two shall be alike but opposite, as an object floating in water is to its reflection, the one containing all the grades above mediocrity up to the highest, and the other all below mediocrity down to the lowest. I do not say that this law is strictly applicable to nations where many individuals are diseased in some definite manner, because the essence of the law is, that the general conditions should be of the same kind throughout. On the other hand, disease and health are for the most part due to little more than different grades of constitutional vigour and of sanitary conditions, and, so far, the nations will fall strictly within the range of the law, which I therefore employ as a useful approximation to the truth. My hope is, that the average standard of a civilised race might be raised to the average standard of the pick of them, as they now are, at the rate of one in every four. It will be clearly understood by those familiar with the law of deviation from an average, that the distribution of ability in a race so improved, would be very different to that of the pick of the present race, though their *average* worth was the same. The improved race would have its broad equatorial belt of mediocrities, and its deviations upwards and downwards, narrowing to delicate cusps; but the vanishing-point of its baseness would not reach so low as at present, and that of its nobleness would reach higher. On the other hand, the pick of our present race would not be symmetrically arranged, but the worst of them would be the most numerous, and the form of the whole body, when classified, would be that of a cone resting on its base, whose sides curved upwards to a sharp point. I find it impossible to explain, without repeating what I have already written, in *Hereditary Genius* (p. 343), the enormous advantages that would follow the elevation of our race through so moderate a range as that I have described. It chiefly consists in the sweeping away of a legion of ineffectives, and in introducing in very much greater proportions the number of men of independent and original thought. It is those men, who form the fine point of the upward cusp, who are the salt of the earth, and who make nations what they are; now the section of the cusp broadens as it descends, therefore if the whole affair be pushed upwards, so to speak, ever so little, the numbers of the men of the same absolute value become very largely increased.

I will endeavour to give an idea of the result of a selection at the rate of 1 in 4 of the inferior specimens of a civilised race, and will take my example from France, because the quality of the nation is well gauged by that of the annual body of youthful conscripts, who are carefully examined, and whose characteristics are minutely classified. It is better not to take too recent a year, as some persons believe the French race to have deteriorated of late, so I will refer to 1859, of which I happen to have the *Compt-rendu sur le Recrutement de l'Armée* in my library. Speaking in round numbers, a quarter of a million of conscripts were examined in that year, and no less than 30 per cent. of that number were rejected as unfit for the army. Six per cent. were too short, being under the puny regulation height of 5 feet 5 inches, and a large proportion of these—say one-half, or 3 per cent.—must be considered as unfit citizens in other respects than being unfitted for the muscular work required in the army. Not many were incapacitated by accident, as by blindness or deafness resulting from injury, or by rupture; but of these, again, only a small portion come justly under that head. I am assured that if a person has hereditary predisposition to deafness, slight accidents, such as a blow on the head, or a bad cold, which would be comparatively harmless to other people, will frequently affect and ruin his hearing; and the same is the case with the eyesight and every other function. In addition, we must recollect that many accidents are the result of stupidity and slowness. Of the injuries by the effects of which youths were unfitted for the army, I feel sure that less than half should be ascribed to pure accident, and that of the 30 per cent. who were rejected for all causes, not more than 3 per cent. should be allowed as coming under that head. Adding this to what we have already excepted out of those who were considered too short, there remain 24 per cent. who were diseased or crippled or puny. In round numbers, one-quarter of the French youths are naturally and hereditarily unfitted for active life.

I will now turn to the other end of the scale of ability, to see what the quarter of a nation is like who are picked out as the best, and I do not know a better example to cite than one which I recently witnessed with great interest; it was on board the *St. Vincent* training ship for seamen for the Royal Navy, which is stationed at Portsmouth. I was informed that out of every three or four applicants not more than one was, on the average, accepted, the applicants themselves being in some degree a selected class. The result was, that when I stood among the 750 boys who composed the crew, it was clear to me that they were decidedly superior to the mass of their countrymen. They showed their inborn superiority by the heartiness of their manner, their self-respect, their healthy looks, their muscular build, the interest they took in what was taught them, and the ease with which they learnt it. A single year's training turns them out accomplished seamen in a large number of particulars. I give in a footnote[2] the conditions which they must fulfil to be qualified for admission; they seem to have been drawn up in an excellent spirit, and to produce most happy results. If the average English youth of the future could be raised by an improvement in our race

---

2 Each boy must bring a proper certificate of character and declaration of age. The age of admission is between 15 and 16. The agreement is to serve in the Navy up to the age of 28. No boys are received from reformatories or prisons, nor if they have been committed before a magistrate. The other requirements are:—

to the average of those on board the *St. Vincent*, which is no preposterous hope, England would become far more noble and powerful than she now is. The general tone of feeling, in short, the 'Mrs. Grundy,' of the nation would be elevated, the present army of ineffectives which clog progress would disappear, and the deviations of individual gifts towards genius would be no less wide or numerous than they now are; but by starting from a higher vantage-ground they would reach proportionately farther.

It is idle to lament the ill condition of our race without bestirring ourselves to find a remedy, but it requires some audacity to publicly propose schemes, because the world at large is incredulous of the extent of the ill, while most of those who are more correctly informed feel little faith in the feasibility of remedying it. Nevertheless, the subject is one which the public ought to be accustomed to hear discussed without surprise or prejudice, and I trust that my own remarks will attract the attention of some few competent persons by whom they may be helpfully criticised. I will describe what I have to propose from the very beginning. It is entirely based on the assumption that the ordinary doctrines of heredity are, in a broad sense, perfectly true; also that the popular mind will gradually become impressed with a conviction of their truth, owing to the future writings and observations of many enquirers; and lastly, that we shall come to think it no hardheartedness to

---

If their age is between	Their height without shoes must be at least	And their measurement round the chest must be at least
15 and $15\frac{1}{2}$	4 feet $10\frac{1}{2}$ inches	29 inches
$15\frac{1}{2}$ and 16	4 feet $11\frac{1}{2}$ inches	$29\frac{1}{2}$ inches
16 and $16\frac{1}{2}$	5 feet 1 inch	30 inches

They must be able to read and write fairly; be strong, healthy, well grown, active, and intelligent; free from all physical malformation; never have had fits, and must be able to pass a strict medical examination by the surgeons of the ship. Their teeth must be good, that they may be able to bite biscuit; at the same time, we must recollect that bad teeth are to some degree the sign of a bad constitution. The applicants come from various directions, and, though a majority of them do not know the regulations for admission, yet, as many of them do, and as all have to bring certificates of character, the applicants, on the average, must be considered to be in some slight degree a selected class.

favour the perpetuation of the stronger, wiser, and more moral races, but shall conceive ourselves to be carrying out the obvious intentions of Nature, by making our social arrangements conducive to the improvement of their race.

There is a vast difference between an intellectual belief in any subject and a living belief which becomes ingrained, sometimes quite suddenly, into the character. I do not venture to ask that the doctrines of heredity shall be popularly accepted in the latter sense, in order that the seeds of my scheme should be planted, but I am satisfied if they shall come to be believed in with about the same degree of persuasion and as little fervour as are those, at the present time, of sanitary science. That is enough to enable the scheme to take root and to grow, but I cannot expect it to flourish until the popular belief shall have waxed several degrees warmer.

My object is to build up, by the mere process of extensive enquiry and publication of results, a sentiment of caste among those who are naturally gifted, and to procure for them, before the system has fairly taken root, such moderate social favour and preference, no more and no less, as would seem reasonable to those who were justly informed of the precise measure of their importance to the nation. I conclude that the natural result of these measures would be to bind them together by a variety of material and social interests and to teach them faith in their future, while I trust to the sentiment of caste to secure that they shall intermarry among themselves about as strictly as is the custom of the nobility in Germany. My proposition certainly is not to begin by breaking up old feelings of social status, but to build up a caste *within* each of the groups into which rank, wealth, and pursuits already divide society, mankind being quite numerous enough to admit of this sub-classification. There are certain ingenious persons who examine the records of unclaimed dividends at the Bank of England, and search for the heirs of the original owners, and inform them (for a consideration) to their advantage. My object is to have the English race explored, and their now unknown wealth of hereditary gifts recorded, and that those who possess such a patrimony should be told of it. I leave it to the natural impulses by which mankind are guided, to insure that such wealth should not continue to be neglected, any more than any other possession unexpectedly made known to them. Great fortunes are commonly observed to coalesce through marriage, and members of aristocracies seldom make alliances out of their order, except to gain wealth. Is it less to be expected that those who become aware that they are endowed with hereditary gifts, should abstain from squandering their patrimony by marrying out of their caste? I do not for a moment contemplate coercion as to whom any given person should marry; such an idea would be scouted now-a-days almost as much as that of polygamy, or of infanticide. But it is quite conformable to the customs of this century to employ social considerations to effect what is desirable, and their efficacy in this case would be as great as is needful. The great majority are sure to yield to it, and it is a trifling matter, when we look to general results, if a small percentage refuse obedience. I also lay great stress on the encouragement of the gifted caste to marry early, and to live under healthy conditions, and this I consider would be effected in the manner I shall briefly explain.

The reader will probably find after I have concluded, that the questions chiefly to be discussed (it being understood that my primary suppositions are provisionally granted) are, first, whether the proposed means are adequate to create a caste whose sentiments shall have the character and strength assigned to them; and secondly, whether the existence of such a caste would or would not be intolerable to the country at large, at the time when it bad become powerful, but by no means dominant.

I propose as the first step, and the time is nearly ripe for it, that some society should undertake three scientific services: the first, by means of a moderate number of influential local agencies, to institute *continuous* enquiries into the facts of human heredity; the second to be a centre of information on heredity for breeders of animals and plants; and the third to discuss and classify the facts that were collected. I look upon the continuity of the enquiry as very important, from the extreme difficulty I have experienced in ransacking bygone family details, even of recent date. Biographies and pedigrees require contemporaneous touching up, in order that they may be full and trustworthy, and that an adequate accumulation of hereditary facts may in time be formed.

All this is purely scientific work, to the performance of which no reasonable objection can possibly be made, and is intended to tell us in what degree and with what qualification the ordinary doctrines of heredity apply to man. Different persons may expect it to yield different results: that which

I expect is, that these doctrines will be fully confirmed in a broad sense, and that an immense amount of supplemental and special information will be gathered. It is entirely on the supposition that these hopes will be verified, that all I have now to say is based. The proposed work is a large one, but not impracticable. Any family or any community could undertake the raw materials for itself, and therefore large districts, or even the entire nation, which is but a collection of such units, could equally do so. However, it would require much enthusiasm in the cause to carry it steadily on, and to discuss the results upon a sufficient scale, but it need not be isolated work. It would naturally fall in with an undertaking that would commend itself to many, of obtaining a more exact statistical insight into the condition of the nation than we now possess, by working very thoroughly a moderate number of typical districts, as samples of our enormous population. If enquirers existed, there are large numbers of statistical queries which might be most usefully answered. Among others, we want an exact stock-taking of our worth as a nation, not roughly clubbed together, rich and poor, in one large whole, but judiciously sorted, by persons who have local knowledge, into classes whose mode of life differs. We want to know all about their respective health and strength and constitutional vigour; to learn the amount of a day's work of men in different occupations; their intellectual capacity, so far as it can be tested at schools; the dying out of certain classes of families, and the rise of others; sanitary questions; and many other allied facts, in order to give a correct idea of the present worth of our race, and means of comparison some years hence of our general progress or retrogression.

I will now suppose a few more years to have passed, during which time short biographies and pedigrees, illustrated by measurements and photographs, shall have been compiled, of perhaps a thousand or more individuals in each of the districts under investigation. Schoolmasters, ministers, medical men, employers of labour, and the resident gentry, will be applied to, but no blind zeal should be evoked that might arouse prejudice and unreasonable opposition. The facts should be collected quietly, and with the *bonâ fide* object of obtaining scientific data. If the results prove to be such as I have reason to expect, then, but only then, will the conviction begin to establish itself in the popular mind, that the influence of heredity is one of extraordinary importance. I ask for no anticipatory action, but merely to enquire on a large scale, in a persistent manner, and to allow events to follow in their natural course, knowing full well that if observation broadly confirms the truth of the present doctrines of heredity, quite as many social influences as are necessary will become directed to obtain the desired end.

I trust that I have made my meaning clear thus far, to the effect that I propose no direct steps at first beyond simple enquiry, but that the mere process of carrying on the enquiries will have an incidental influence in creating common interests and mutual acquaintance and friendships among the gifted families in each class of society, such effects naturally resulting in frequent cases of intermarriage. Then I say, the offspring of these intermarriages will have some moderate claim to purity of blood, because their parents and many of their more distant relatives will be gifted above the average; also, the precise family history of each of them will have been preserved, and the foundation laid of a future 'golden book' of natural nobility. Lastly, a mass of information bearing on human heredity will have been collected.

In the meantime (supposing the fundamental truth of all I maintain as regards the doctrine of heredity, and the probability that the improvement of the human race will be considered a duty) the scale on which enquiries are conducted will steadily grow. I should expect that all boys at school will not only be examined and classed, as at present, for their intellectual acquirements, but will be weighed and measured and appraised in respect of their natural gifts, physical and mental together, and that enquiries will, as a matter of course, be made into the genealogies of those among them who were hereditarily remarkable, so that all the most promising individuals in a large part of the kingdom would be registered, each in his own local centre. A vast deal of work would be, no doubt, thrown away in collecting materials about persons who afterwards proved not to be the parents of gifted children. Also many would be registered on grounds which our future knowledge will pronounce inadequate. But gradually, notwithstanding many mistakes at first, much ridicule and misunderstanding, and not a little blind hostility, people will confess that the scheme is very reasonable, and works well of its own accord. An immense deal of investigation and criticism will bear its proper fruit, and the cardinal rules for its successful procedure will become understood and laid down. Such, for example, as the physical, moral, and intellectual qualifications for entry on the register, and especially as to the increased importance of those which are not isolated, but common to many members of the same family. It will be necessary also to have a clear idea of the average order of gifts to aim for, in the race of the immediate future, bearing in mind that sudden and ambitious attempts are sure to lead to disappointment. And again, the degree of rigour of selection necessary among the parents to insure that their children should, on the average, inherit gifts of the order aimed at. Lastly, we should learn particulars concerning specific types, how far they clash together or are mutually helpful.

Let us now suppose an intermediate stage to be reached, between that of mere investigation and that of an accepted system and practical action, and try to imagine what would occur. The society of which I have been speaking, or others like it, would continually watch the career of the persons whose names were on their register, and those who had aroused so much interest would feel themselves associates of a great guild. They would be accustomed to be treated with more respect and consideration than others whose parents were originally of the same social rank. It would be impertinent in anyone to assume airs of patronage towards such people; on the contrary, the consideration shown them would naturally tend to encourage their self-respect and the feeling that they had a family name to support and to hand down to their descendants. Again, the society would be ever watchful and able to befriend them, for it would be no slight help to a man to state, on undoubted grounds, that not only is he what he appears, but that he has latent gifts as well. That he is likely to have a healthy life, and that his children are very likely indeed to prove better than those of other people. In short, that he and his family may be expected to turn out yet more creditably than those ignorant of his and his wife's hereditary gifts would imagine. This would make it more easy for him than for others to obtain a settled home and employment in early manhood, and to follow his natural instinct of marrying young. It is no new thing that associations should successfully watch and befriend every member of large communities, and in the present case the kindly interests sure to be evoked in dealing with really worthy and self-helpful people would be so great that I should expect charity of this kind to become exceedingly popular, and to occupy a large part of the leisure of many people. It is quite another thing to patronising paupers, and doing what are commonly spoken of as 'charitable' actions, which, however devoted they may be to a holy cause, have a notorious tendency to demoralise the recipient, and to increase the extent of the very evils which they are intended to cure.

The obvious question arises, Would not these selected people become intolerably priggish and supercilious? Also it will be said, that the democratic feeling is a growing one, and would be directly adverse to the establishment of such a favoured and exceptional class. My answer is, that the individuals in question would not at first have so very much to be conceited about, and that, later on, their value would be generally recognised. They would be good all round, in physique and *morale*, rather than exceptionally brilliant, for many of the geniuses would not 'pass' for physical qualities, and they would be kept in good order by the consciousness that any absurd airs on their part might be dangerous to them. The attitude of mind which I should expect to predominate, would be akin to that now held by and towards the possessors of ancestral property, of moderate value, dearly cherished, and having duties attached. Such a person would feel it a point of honour never to alienate the old place, and he is generally respected for his feeling and liked on his own account. So a man of good race would feel that marriage out of his caste would tarnish his blood, and his sentiments would be sympathised with by all. As regards the democratic feeling, its assertion of equality is deserving of the highest admiration so far as it demands equal consideration for the feelings of all, just in the same way as their rights are equally maintained by the law. But it goes farther than this, for it asserts that men are of equal value as social units, equally capable of voting, and the rest. This feeling is undeniably wrong and cannot last. I therefore do not hesitate in believing that if the persons on the register were obviously better and finer pieces of manhood in every respect than other men, democracy notwithstanding, their superiority would be recognised at just what it amounted to, without envy, but very possibly with some feeling of hostility on the part of beaten competitors.

Let us now, in our imagination, advance a couple of generations, and suppose a yet more distant time to have arrived, when societies shall have been sown broadcast over the land and have become firmly rooted, and when principles of selection shall have been well discussed and pretty generally established, and when, perhaps, one per cent. of the thirty millions of British people, that is 300,000 individuals, old and young, and of both sexes, shall have their names inserted in the then annually published registers. By this time the selected race will have become a power, a considerable increase will have taken place in the number of families of really good breed, for there will be many boys and girls, themselves above mediocrity, whose parents, uncles on both sides, four grandparents, several of their great-uncles and cousins, and all their eight great-grandparents, were persons considerably above the average in every respect that fits an individual to be a worthy citizen and a useful and agreeable member of society. I cannot doubt, that at this period a strong feeling of caste would be found developed in the rising generation, for such is the vanity of men, especially in youth, that it is one of the easiest tasks in the world to persuade them that they are in some way remarkable, and, in the supposed case, the persuasion would be well-nigh irresistible. A number of, perhaps, the best informed philosophers in the nation, who are experts in the matter, solemnly aver, after careful enquiry, that the individuals whose names are on the register are, in sober truth, the most valuable boys and girls, or men and women, to the nation. They may give them a diploma, which would virtually be a patent of natural nobility. They assure them that if they intermarry under certain limitations of type and subclass, which have yet to be studied and framed, their children will be, on the whole, better in every respect than the children of other people—stronger, healthier, brighter, more honest and more pleasant. They tell them that in addition to the old-established considerations of rank and wealth there is another and a higher one, namely, of purity of blood, and that it would be base to ally themselves with inferior breeds. In corroboration of these flattering words, the members of the gifted caste would continue to experience pleasing testimony of a practical kind, for there can be little doubt that one consequence of the continual writing and talking about noble races of men, during many years, would be to increase the appreciation of them. An entry on the register would then become as beneficial as it was a few years since to be born of a family able and willing to push forward their relatives in public life. Queen Elizabeth gave ready promotion to well-made men, and it is no unreasonable expectation that our future landowners may feel great pride in being surrounded by a tenantry of magnificent specimens of manhood and womanhood, mentally and physically, and that they would compete with one another to attract and locate in their neighbourhood a population of registered families.

I will now suppose another not improbable alternative, namely, the result of some democratic hostility to the favoured race. Well, it would gain in cohesion by persecution. If trade unionism chose to look on them as cuckoos in the national nest, they would be driven from the workshops, and be powerfully directed to cooperative pursuits. They would certainly have little inclination to inhabit towns where they were outnumbered and disfavoured, and would naturally settle in co-operative associations in the country. In other words, the gifted race would be urged into companionship by the pressure of external circumstances, no less strongly than, as I have shown, they would be drawn together by their own mutual attraction, and would be perforce inhabitants of healthy rural districts, and not of unhealthy towns. All this, which is probable enough, would have an immense effect in strengthening the sentiment of caste, in developing the best points of their race, and in increasing its numbers. In these colonies, caste regulations would no doubt rise into existence, and gradually acquire the force almost of religious obligations, to maintain and increase the character of their race, by encouraging early marriage among their more gifted descendants, and by discouraging it among the less gifted. The colonies would become more and more independent as the superiority of their members over the outside world became, in successive generations, more pronounced. Their members would be little likely to associate intimately with persons not of their caste, because they would succeed better by themselves than when other and less effective men were admitted into partnership. They would not only have peculiarly high personal gifts of intelligence and *morale* to carry out co-operative undertakings, but they would also have in many cases special advantages as well.

If they wished to found a club for mutual relief in sickness, it would be foolish to allow strangers of a less healthy race to join with them. If it should be a building society, they by themselves would be able to enforce better sanitary regulations than if a body of less intelligent and energetic families were mixed up with them. Their social gatherings would tend to be exclusive, because their interests would be different, and often hostile, to those of other people, and their own society would be by far the more cultured and pleasant.

It will be understood that the colonies I am describing, would be large enough for all the varied interests of life to find place for their exercise. They would be no mere retreats from a distasteful outside world, but energetic and capable to the higher degree. The continued intermarriage of members of such colonies seems to me almost a certainty, and so does the happiness which would generally be diffused among them. Here, if anywhere, would a whole population learn to be industrious, like bees or ants, for public ends and not for individual gain. If such communities were established, it would be in them, rather than anywhere else, where those forms of new and higher civilisation, which must hereafter overspread the earth, would be first evolved. If, however, they should be persecuted to an unreasonable extent, as so many able sects have already been, let them take ship and emigrate and become the parents of a new state, with a glorious future.

All I have thus far spoken of would require no endowments, and yet how much could be effected by it. We may, however, expect that endowments commensurate with the greater items of national expenditure would ultimately be assigned to the maintenance and improvement of the best races of man. Our peers enjoy a gross annual income of some nine millions; and that of all other settled property, irrespective of merit, would amount to an enormous sum. It is very possible hereafter, at the time I have been anticipating, that the Legislature under the growing influence of the gifted caste (supposing other customs to remain as they are at present) would enforce some limitation to inheritance, in cases where the heirs were deficient in natural gifts. The fittest would then have a far better chance of survival than at present, and civilisation, which is now recklessly destructive of high races, would, under more enlightened leadership, employ its force to maintain and improve them. The gifted families would be full of life and hope, and living under more intelligent and favourable sanitary conditions, would multiply rapidly, while the nongifted would begin to decay out of the land, whenever they were brought face to face in competition with them, just in the same way as inferior races always disappear before superior ones. It is difficult to analyse the steps by which this invariable law has hitherto accomplished itself, and much more difficult is it to guess how it would be accomplished under the conditions here described, but I should expect it would be effected with little severity. I do not see why any insolence of caste should prevent the gifted class, when they had the power, from treating their compatriots with all kindness, so long as they maintained celibacy. But if these continued to procreate children, inferior in moral, intellectual and physical qualities, it is easy to believe the time may come when such persons would be considered as enemies to the State, and to have forfeited all claims to kindness.

The objection is sure to be urged against my scheme, that its effects are too remote for men to care to trouble themselves about it. The earlier results will be insignificant in number, and disappointing to the sanguine and ignorant, who may expect a high race to be evolved out of the present mongrel mass of mankind in a single generation. Of course this is absurd; there will be numerous and most annoying cases of reversion in the first and even in the second generation, but when the third generation of selected men has been reached, the race will begin to bear offspring of distinctly purer blood than in the first, and after five or six generations, reversion to an inferior type will be rare. But is not that too remote an event for us to care for? I reply that the current interests which the scheme would evoke are, as already explained, of a very attractive kind, and a sufficient reward for considerable exertion quite independently of anything else. Its effects would be ever present, clearly visible, of general importance, and of the highest interest, the number of experiments going on at the same time being an equivalent to the slowness with which their results became apparent. Also, it must be recollected that the labourers employed on the foundation of any edifice, have a store of present pleasure in discounting, so to speak, its future development.

But even if the labour were wholly unremunerated by present pleasure, I should not despair, looking at the great works already accomplished under similar conditions. I will cite one example. The forests of Europe[3] extend over enormous tracts. In France, alone, they cover between eight and nine million acres, which equals a region 130 miles long by 100 broad. The chief timber tree in France is oak, and an ordinance which

---

[3] I take all the following facts from a very curious and interesting memoir by Mr. Sykes Gamble, Assistant Conservator of Forests in British India, published in the *Transactions of the Highland and Agricultural Society of Scotland*, 1872.

dates from 1669 contains a clause inserted by Colbert that in none of the forests of the State shall oaks be felled until they are ripe, that is, are unable to prosper for more than thirty years longer. This regulation has been strictly attended to up to the present day, and in the mean time forest legislation has grown into an important duty of the State. The same has occurred in Germany, and the lead of these two countries has been followed by Italy, Prussia, Austria, Sweden, Denmark, and British India. To return to our oaks: the timber is of great value in France, not only for ship building, but on account of the enormous quantity used for parquet floors and wine casks, while, on the other hand, countries which formerly supplied it in abundance, are now running short. In North Germany oaks are rarely permitted to attain a large size, being usually felled before they are 100 years of age, and the fine natural forests of Hungary, Croatia and Sclavonia are becoming exhausted; consequently the Government of France strives to favour in every way the growth of fine oak timber and postpones felling the trees until they are fully mature; that is, between the ages of 150 and 180 years.

Is not man worthy of more consideration than timber? If a nation readily consents to lay costly plans for results not to be attained until five generations of men shall have passed away, for a good supply of oak, could it not be persuaded to do at least as much for a good supply of man? Marvellous effects might be produced in five generations (or in 166 years, allowing three generations to a century). I believe, when the truth of heredity as respects man shall have become firmly established and clearly understood, that instead of a sluggish regard being shown towards a practical application of their knowledge, it is much more likely that a perfect enthusiasm for improving the race might develop itself among the educated classes.

Chapter 12

Africa for The Chinese[1]

Sir,—In a few days Sir Bartle Frere will return to England, and public attention will be directed to the East Coast of Africa.[2] I am desirous of availing myself of the opportunity to ventilate some speculations of my own, which you may, perhaps, consider of sufficient interest to deserve publication in the *Times*. My proposal is to make the encouragement of the Chinese settlements at one or more suitable places on the East Coast of Africa a part of our national policy, in the belief that the Chinese immigrants would not only maintain their position, but that they would multiply and their descendants supplant the inferior Negro race. I should expect the large part of the African seaboard, now sparsely occupied by lazy, palavering savages living under the nominal sovereignty of the Zanzibar, or Portugal, might in a few years be tenanted by industrious, order loving Chinese, living either as a semi-detached dependency of China, or else in perfect freedom under their own law. In the latter case their position would be similar to that of the inhabitants of Liberia, in West Africa, the territory which was purchased 50 years ago and set apart as an independent State for the reception of freed negroes from America.

The opinion of the public on the real worth of the Negro race has halted between the extreme views which have been long and loudly proclaimed. It refuses to follow those of the early abolitionists, that all the barbarities in Africa are to be traced to the effects of a foreign slave trade, because travellers continually speak of similar barbarities existing in regions to which the slave trade has not penetrated. Captain Colomb

[1] Galton 1873a.

[2] Frere, Sir Bartle Edward, see page 356. Frere had returned from his anti-slavery mission to Zanzibar.

has written a well-argued chapter on this matter, in his recent volume.[3] On the other hand, the opinion of the present day repudiates the belief that the negro is an extremely inferior being, because there are notorious instances of negroes possessing high intelligence and culture, some of whom acquire large fortunes in commerce, and others become considerable men in other walks of life. The truth appears to be that individuals of the mental calibre I have just described are much more exceptional in the negro than in the Anglo-Saxon race, and that average negroes possess too little intellect, self-reliance, and self-control to make it possible for them to sustain the burden of any respectable form of civilization without a large measure of external guidance and support. The Chinaman is a being of another kind, who is endowed with a remarkable aptitude for a high material civilization. He is seen to the least advantage in his own country, where a temporary dark age still prevails, which has not sapped the genius of the race, though it has stunted the development of each member of it, by the rigid enforcement of an effete system of classical education which treats originality as a social crime. All the bad parts of his character, as his lying and servility, spring from timidity due to an education that has cowed him, and no treatment is better calculated to remedy that evil than location in a free settlement. The natural capacity of the Chinaman shows itself by the success with which, notwithstanding his timidity, he competes with strangers, wherever he may reside. The Chinese emigrants possess an extraordinary instinct for political and social organization; they contrive to establish for themselves a police and internal government, and they give no trouble to their rulers so long as they are left to manage those matters by themselves. They are good-tempered, frugal, industrious, saving, commercially inclined, and extraordinarily prolific. They thrive in all countries, the natives of the Southern provinces being perfectly able to labor and multiply in the hottest climates. Of all known varieties of mankind there is none so appropriate as the Chinaman to become the future occupant of the enormous regions which lie between the tropics, whose extent is far more vast than it appears, from the cramped manner in which those latitudes are pictured in the ordinary maps of the world. But take a globe and examine it, and consider the huge but poorly-peopled bulk of Africa, by

---

[3] Philip Howard Colomb *Slave catching in the Indian Ocean: a record of naval experiences* (London: Dawsons, 1873).

whose side the areas of India and of China look insignificant, and think what a field lies there for the development of a suitable race. The Hindoo cannot fulfil the required conditions nearly as well as the Chinaman, for he is inferior to him in strength, industry, aptitude for saving, business habits, and prolific power. The Arab is little more than an eater up of other men's produce; he is a destroyer rather than a creator, and he is unprolific.

The history of the world tells a tale of the continual displacement of populations, each by a worthier successor, and humanity gains thereby. We ourselves are no descendents of the aborigines of Britain, and our colonists were invaders of the regions they now occupy as their lawful home. But the countries into which the Anglo-Saxon race can be transfused are restricted to those where the climate is temperate. The Tropics are not for us, to inhabit permanently; the greater part of Africa is the heritage of people differently constituted to ourselves. On that continent, as elsewhere, one population continually drives out another. Consider its history as it extends over successive centuries. We note how Arab, Tuarick, Fellatah, Negroes of uncounted varieties, Caffre, Hottentot surge and reel to and fro in the struggle for existence. It is into this free flight among all present that I wish to see a new competitor introduced—namely, the Chinaman. The gain would be immense to the whole civilized world if we were to out-breed and finally displace the negro, as completely as the latter has displaced the aborigines of the West Indies. The magnitude of the gain may be partly estimated by making the converse supposition—namely, the loss that would ensue if China were somehow to be depopulated and restocked by negroes.

The pressure of population in China is enormous, and its outflow is great and increasing. There is no lack of material for a suitable immigration into Africa. I do not say that it would be possible at any moment to persuade communities of men and women from Southern China to establish themselves in Africa; but I am assured, by excellent authorities, that occasions of political disturbances frequently arise when it would be practicable to do so by the promise of a free, or nearly free, grant of land. The Chinese have a land hunger, as well as a love for petty traffic, and they would find a field in which to gratify both of these tastes on the East African Coast. There are many Chinese capitalists resident in foreign parts who might speculate in such a system and warmly encourage it. If once successfully started, it ought to maintain itself. The colonist could not starve; and when they began to succeed they would send money to their relatives to enable them to follow, just as they now do from the many other parts of the world where they are located. For these reasons it is probable that the streams of emigration from China have sufficient "head" to enable them to reach and overflow the coasts of Eastern Africa if they were watched and judiciously diverted in that direction.

I have finally to speak of the political effort necessary to secure a free right of occupancy and of settlement at suitable points on the coast. No very serious obstacle seems to stand in the way; certainly none was met with when Liberia was founded. It is probable that as the success of such an enterprise would be of equally great value to all nations commercially interested in those parts, no national jealousy would be excited by its promotion, and the necessary territory could be obtained with little difficulty and at a small cost, to be advanced in the first instance as a charge on the land and hereafter to be redeemed.

Francis Galton

Chapter 13

On Men of Science, their Nature and their Nurture.

WEEKLY EVENING MEETING, Friday, February 27, 1874. George Busk, Esq. F.R.S. Treasurer and Vice-President, in the Chair.[1]

Francis Galton, Esq. F.R.S. M.R.I. On Men of Science, their Nature and their Nurture.

The purport of this discourse is to specify the chief qualities by which the English men of science of the present day are characterized, to show the possibility of defining and roughly measuring the amount of any of those qualities, and to conclude by summarizing the opinions of the scientific men on the merits and demerits of their own education, giving an interpretation of what, according to their own showing, they would have preferred. My data are obtained from a large collection of autobiographical notes, most obligingly communicated to me by a large part of the leading members of the scientific world. Applications were addressed to 180 Fellows of the Royal Society, who, in addition to their "F.R.S.," had gained medals or filled posts of recognized scientific position; 115 answers have already been received, of which 80 or [90][2] are full and minute replies to my long and varied series of questions. But I can deal with only a few deductions from this valuable material, and must refer to a forthcoming work for the rest.

It is of interest to know the ratio which the numbers of the leading scientific men bear to the population of England generally. I obtain it in this way. Although 180 persons only

[1] Galton 1874b

[2] Here 9 appears but is surely a misprint.

were on my list, I reckon that it would have been possible to have included 300 of the same ages, without descending in the scale of scientific position; also it appears that the ages of half of the number on my list lie between 50 and 65, and that about three-quarters of these may be considered English. I combine these numbers, and compare them with that of the male population of England and Wales, between the same limits of age, and find the required ratio to be about one in 10,000.

What then are the conditions of nature, and the various circumstances and conditions of life,—which I include under the general name of nurture,—which have selected that one and left the remainder? Some may feel surprise that so many as 300 persons are to be found in the United Kingdom who deserve the title of scientific men; probably they have been accustomed to concentrate their attention upon a few notabilities, and to ignore their colleagues. It must, however, be recollected that all biographies, even of the greatest men, reveal numerous associates and competitors whose merit and influence were far greater than had been suspected by the outside world. Great discoveries have often been made simultaneously by workers ignorant of each other's labours. This shows that they had derived their inspiration from a common but hidden source, as no mere chance would account for simultaneous discovery. It would appear that few discoveries are wholly due to a single man, but rather that vague and imperfect ideas, which float in conversation and literature, must grow, gather, and develop, until some more perspicuous and prompt mind than the rest, clearly sees them. The first discoverers beat their contemporaries in point of time, and it is therefore due to them, not that science progresses, but that her progress is as rapid as it is. We must neither underrate nor overrate their achievements. I would compare the small band of men who have achieved a conspicuous scientific position, to islands, which are not the detached phenomena they appear, but only the uppermost portions of hills, whose bulk is unseen. To pursue this metaphor; the range of my inquiry dips a few fathoms below the level at which popular reputation begins.

I proceed to speak of the qualities which the returns specify as most conspicuous in scientific men, and I shall endeavour to make them tell their own tale by quoting anonymous extracts from their communications. The first in order of importance is energy, both of body and of mind. It appears to be possessed in an unusual degree by three-fourths of the men in my list. I should mention that the list contains a very few names of travellers of extraordinary endurance, such as the late Dr. Livingstone; but that I do not speak of these in the following extracts:— 1. "Have rowed myself in a skiff 105 miles in 21 hours whilst undergraduate at Cambridge." (This is, I believe, a feat that not one undergraduate in 500 could do.) " Rowed in every race during my stay at the University; rowed two years in the University crews." 2. "Walked many a time 50 miles a day without fatigue, and kept up five miles an hour for three or four hours." 3. "Excelled at school and college in athletic sports, especially in jumping (18 feet). Almost incapable of mental fatigue up to the age of 38. Usually engaged in literary work until long after midnight." 4. "As a boy of 17, I worked for three months a day and all night with not more than four or five hours sleep. When full of a subject and interested in it, I have written for seven or eight hours without interruption."

Severe scientific work is often done during the night by men engaged all day in anxious business; thus:—"In early life as a boy, I was engaged in business from twelve to fourteen hours a day, yet always found time to study and make my own instruments. Later on, my studies and scientific work were always accomplished after business hours, and it was generally my habit to commence after dinner, and to work at science until 2, 3, or 4 a.m., and to begin business again at 9 a.m. I never thought of rest if I had anything in hand of interest."

I may mention that energy appears to be correlated with smallness of head, a fact which comes out conspicuously here, although the average circumference of head among the scientific men is great. Energy is also, as we have seen, strongly marked among them; but it is much the more strongly marked among those who have small heads. I have ninety-nine returns, many of which I have verified myself, using the hatmaker's whalebone-hoop and measuring inside the hats. It appears that the average circumference of an English gentleman's head is $22\frac{1}{4}$ to $22\frac{1}{2}$ inches. Now, I have only thirteen cases under 22 inches and eight cases of 24 inches or upwards. The general scientific position of the small-headed and large-headed men seems equally good; but the fact is conspicuous that, out of the thirteen of the former, there are only two or three who have not remarkable energy; and out of eight of the latter there is only one who has. A combination of great energy and great intellectual capacity is the most effective of all conditions; but, like the combination of swiftness and strength in muscular powers, it is very rare.

The excellence of the health of the men in my list is remarkable, considering that the majority are of middle and many of advanced ages. One quarter of the whole have excellent or very good health, a second quarter have good or fair, a third have had good health since they attained manhood, and only one quarter make complaints or reservations. Here are two examples of excellent health:— 1. "Only absent from professional duties two days in thirty years; only two headaches in my life." The next is from a correspondent who is between 70 and 80 years of age. 2. "Never ill for more than two or three days except with neuralgia; no surgical operations except inoculation, drawing of one tooth, and cutting of corns." It is positively startling to observe in these returns the strongly hereditary character of good and indifferent constitutions. I have classified the entries, each entry giving the health of the scientific man, of his father and of his mother respectively, and find as follows: —First, a long row of such terms as these: "Excellent; excellent; excellent;" or "Good; good; good" then comes another row in which some ailment is specified by the scientific man as affecting himself, and as having also affected one or other of his parents.

Examples:— 1. "Excellent, but hay fever; father, excellent, but severe hay fever." 2. "Good in early life, subject to headache; father, good, subject to headache." 3. "Delicate in early life, one lung seriously affected; mother delicate and phthisical." I can find only two cases, neither very strongly marked, in which both parents were described as unhealthy, although marriages between such persons are not infrequent. These returns seem to show that the issue of such marriages are barely capable of pushing their way to the front ranks of life. All statistical data concur in proving that healthy persons are far more likely than others to have healthy progeny; and this truth cannot be too often illustrated, until it has taken such hold of the popular mind, that considerations of health and energy shall be of recognized importance in questions of marriage, as much so as the more immediately obvious ones of rank and fortune. Steady perseverance is a third quality on which much stress is laid, but this might have been anticipated, and it is unnecessary to quote instances.

Some prevalence of practical business habits might also

have been anticipated, but it proves much more common than I had expected. Among those who have sent me returns, I count no less than seventeen who are active heads of great commercial undertakings. There are also ten medical men in the highest rank of practice, and eighteen others who fill or have filled important official posts. A most eminent biologist wrote as follows, in reply to the inquiry whether he had any special tastes bearing on scientific success, in addition to those for his own line of investigation:—"I have no special talent except for business, as evinced by keeping accounts, being regular in correspondence, and investing money very well." It is clear that method and order are essential to the man who hopes to deal successfully with masses of details.

Next, as regards the more special qualities; those already mentioned, of energy, health, steadiness of pursuit, and business habits being of general application. The first of these is independence of character. Fifty of my correspondents show that they possess it in excess, and in only two is it below par. Here are a few examples:— 1. "Left aet. 12" [that is, ran away from] "a school where I had received injustice from the master." 2. "Opinions in almost all respects opposed to those in which I was educated." 3. "I have always taken my own independent line. My heresy prevented my advancement." 4. "Preference for whatever is not the fashion, not popular, not rich, not very able to help itself, yet with qualities unworthily overlooked or unjustly oppressed."

The home atmosphere which the scientific men breathed in their youth was generally saturated with the spirit of independence. Examples:— 1. "My father was extremely independent, in some respects more so than I am. He never took off his hat to anyone in his life, and never addressed anyone as Esq." 2. "My father was a Liberal when Liberalism (then styled Jacobinism) was highly obnoxious, an early denouncer of slavery and advocate of religious liberty, a free trader when the world was protectionist, and an opponent of unrighteous war when war was most popular. He was for mitigating our criminal code when hanging was regarded as the sheet-anchor, and, in a word, was politically and socially a very independent spirit."

In confirmation of the assertion that the scientific men were usually brought up in families characterized by independence of disposition, I would refer to the strange variety of small and unfashionable religious sects to which they or their parents belonged. We all know that Dalton, the discoverer of the atomic theory, and Dr. Young, of the undulatory theory of light, were both Quakers, and that Faraday was a Sandemanian. So I find in these returns numerous cases of Quaker pedigree; and I know of one man, not as yet technically on my list, who was born a Sandemanian. There are also representatives of several other small sects, as Moravians and Bible Christians, and the Unitarians are numerous. It will be understood that the object of saying this is not to throw light on the religious tendencies of the scientific men (concerning which I have much material), because so off-hand a statement would mislead, but to prove that they and their parents had the habit of doing what they preferred, without considering the fashion of the day. The man of science is thoroughly independent in character.

We now come to what I look upon as the salt of the character of most scientific men, namely, strong innate taste for science or for some special branch of it. It is not universal even among those who have had the highest success, but it is very common, and it sometimes attains to the height of a passion which is not transient, but abides. Though decidedly hereditary in numerous cases, its appearance is more capricious than health or energy, and it often happens that the scientific man is the only member of his family in whom the taste has shown itself. The following are a few examples of innate taste:— 1. "Thoroughly innate; I had no regular instruction, and can think of no event which especially helped to develop it. Bones and shells were attractive to me before I could consider them with apparent profit, and I had a fair zoological collection by the time I was 15." 2. "If any tastes be innate, mine were. They date from beyond my recollection. They were not determined by events occurring after manhood, but I think the reverse; they were discouraged in every way." 3. "I should say innate. As to whether they were largely determined by events occurring after manhood, I think not. All I can say is, that neither profession, nor marriage, nor sickness, have been able to affect them." 4. "As far back as I can remember, I loved nature and desired to learn her secrets, and I have spent my whole life in searching for them. While a schoolboy I taught myself ... under great difficulties."

Let us now put these results together. We have seen that energy, health, steady pursuit of purpose, business habits, independence of character, and a strong innate taste for science, are characteristics of scientific men. Probably one half of the men on my list possess every one of them in a considerable and some of them in a very high degree. If one or more of these qualities be deficient, success becomes impossible, unless its absence is supplemented by other and as yet unclassified conditions. The want of time prevents me from entering into these, and I must postpone further results to a future publication. However, two groups of cases may be specified in which only a few of the above-mentioned qualities are present, and which end in an abortive career. The one is the possession of energy, health, and independence of character in excess, and little else to control them.

These are dangerous gifts. Those who have them are apt to renounce guidances by which the great body of mankind move safely, and to follow out a career in which they are almost certain to blunder and fail egregiously. Probably every large emigrant ship takes out many such men, full of unjustifiable self-confidence, who, to use a current phrase, "knock about in the world," waste their health, youth, and opportunities, and end broken down. Another common group of cases are those where a strong innate taste for science is accompanied by independence of character and steadiness of pursuit, but with no other quality helpful to success, and who therefore fail. There is hardly a village where some ingenious man may not be found who has ideas and much shrewdness, but is crotchety and impracticable. He wants energy and business habits, and so he never rises. There are many who brood over subjects like perpetual motion, whose peculiarities are well illustrated in De Morgan's book of paradoxes. We also frequently meet persons of the stamp that justifies the old-fashioned caricature of scientific men, being absorbed in some petty investigation, utterly deficient in business habits, and noted for absence of mind. I may add that even idiots have often strongly quasi-scientific tastes, as love for simple mechanism, or objects of natural history, and they have a pleasure in collecting. Also, we all know that madmen have often persistency, as shown by their brooding on a single topic.

Lastly, I wish to give some idea of the very general prevalence of mechanical tastes among the scientific men generally. One would have expected to find it among mechanicians and physicists, but it is just as strong among the biologists and others. One chemist made a 12-inch reflecting telescope; two eminent surgeons have an extraordinary aptitude for and love of mechanical manipulation; two very eminent biologists had a passion for it, and both, if they had followed the bent of their own minds, would have been engineers by profession.

All tends to show that the scientific mind is directed to facts and abstract theories, and not to persons or human interests. The man of science is deficient in the purely emotional element, and in the desire to influence the beliefs of others. Thus I find that two out of every ten do not care for politics at all; they are devoid of partisanship. They school a naturally equable and independent mind to a still more complete subordination to their judgment. In many respects their character is strongly anti-feminine. It is a curious proof of this, that in the very numerous answers which have reference to parental influence, that of the father is quoted three times as often as that of the mother. It would not have been the case, judging from inquiries I elsewhere made, if I had been discussing literary men, commanders, or statesmen, or, still more, divines.

I regret much that time makes it impossible for me now to dive deeper into the rich mine of facts contained in my returns. It becomes necessary for me to leave this branch of the subject and to pass on to some interesting considerations regarding the measurement of qualities such as those we have been engaged upon. These considerations are of the most general application, and are as applicable to magnitude as they are to intellect and morals, and to every form of animal or vegetable life as they are to men. I shall therefore speak about the size of nuts, and peas, and acorns, as being easily experimented on, and deduce from these the results which I would fain apply to the moral and intellectual qualities of mankind.

The law of statistical constancy may be taken for granted. It is evidenced by the experience of insurance offices against fire, death, shipwreck, and other contingencies, always with the proviso that the facts are gathered with discretion, on well-known general principles. Hence we may say with assurance, that although two common nuts may differ, yet the contents of different packets, each containing 1000 nuts, will be scarcely distinguishable, for the same number of nuts of different sizes will be found in each. Let the contents of the several packets be each arranged in a long row, in order of size, beginning with the biggest nut and ending with the smallest, and place the rows rank behind rank; then by the law of statistical constancy the nuts in the same files will in all cases be closely alike (except the outside ones, where more irregularity prevails). Again, if we incorporate two rows into one of double length, still preserving the arrangement as to regular gradation in size, the centre nuts of the two original series will still be found at or near the centre of the compound series, the nuts in quarter positions will still be in quarter positions, and so on. Hence, whatever be the length of the series the *relative position* in it of the nut will be a *strict criterion* of its size. This is of course equally true of all groups of qualities or characters whatever, in which the law of statistical constancy prevails, the series, in each case, being arranged according to gradations of the quality in question. Each individual is measured against his neighbour, and it is quite unnecessary to have recourse to any external standard. As regards a scale of equal parts, I make use of a converse application of the law of "frequency of error" [this was illustrated by many experiments], which shows that in a row (say as before) of nuts, if we take those which occupy the three quarterly divisions (1st quarter, centre, 3rd quarter) as three elementary graduations of size, a continuous scale of graduations will be determined by the following series, in which the places of the nuts are supposed to be reckoned from the end of the row where the large nuts are situated, and to be given in per-thousandths of the entire length of the row. It might be called the "Common Statistical Scale" (S. S.). The place of +4° would be at 4 thousandths from [the] large end; +3° at 21 thousandths; +2° at 89; +1° at 250; 0° at 500; -1° at 750; -2° at 911; -3° at 979; and -4° at 996, or 4 thousandths from the small end of the row.

Thus if we say that the size of a nut is + 2°S. S., we absolutely define, or rather identify, what we are speaking about. Anybody can procure such a nut independently by getting a quart of nuts and arranging them. Also we know that the difference between a nut of +4°S.S. and + 1°S.S. is 3°, and therefore three times as great as between one of + 2°S.S. and the latter. It cannot be affirmed that this is a precise scale of equal parts for all qualities, but it is found to hold surprisingly well in a great variety of vital statistics; perhaps, too, the mere thickness of tissues may be a chief element in the physical basis of life. This scale appears, at all events, more likely to be nearly approximative to one of equal parts, for qualities generally, than any other that can be specified, and it certainly affords definite standards subject to the law of statistical constancy. The habit should therefore be encouraged in biographies, of ranking a man among his contemporaries, in respect to every quality that is discussed, and to give ample data in justification of the rank assigned to him. By the general use of a system like the above, which is universally applicable, social and political science would be greatly raised in precision.

I now pass on to the education which the scientific men had in their youth, in the hope that my results may give assistance to those who are endeavouring to frame systems of education suitable to the wants of the day. What I have to say, is very partial; it refers solely to the opinions the scientific men entertain of the merits and faults of their own several educations. Their views are remarkably unanimous, considering the very different branches of inquiry they are interested in, and the great dissimilarities in their education. I should mention, that one-third have been educated at Oxford or Cambridge, one-third at Scotch, Irish, or London Universities, and the remaining third have been at no University at all. I am totally unable to decide which of the three groups occupies the highest scientific positions, they seem to me very much alike in this respect. The merits they all ascribe to variety of education are to be gathered from the following examples:— 1. "Not tied down to old courses of "classics and mathematics." 2. "Sufficient groundwork in many subjects to avoid error." 3. "Early introduced to many subjects of interest." 4. "A well-balanced education, including chemistry, botany, logic, and political economy." 5. "A variety of subjects, and attention to details." 6. "Coming in contact with persons of every rank and sitting in the same form "[in a Scotch school]" with the sons of tradesmen, and ploughmen, as well as gentlemen."

In contrast to this, here are some examples as to the faults of their education:— 1. "No mathematics, nor modern languages, nor any habits of observation or reasoning." 2. "Enormous time devoted to Latin and Greek, with which languages I am not conversant." 3. "Omission of almost everything useful and good, except being taught to read; Latin, Latin, Latin!" 4. "In an otherwise well-balanced education, three years were spent on Latin and Greek grammar, a blank waste of time." 5. "Neglect of many subjects for the attainment of one or two; not pushing mathematics to a useful end."

Evidence such as this, which could be largely added to, establishes the advantage of variety of study. One group of men speak gratefully because they had it, and another group speak regretfully because they had it not. I find none who had a reasonable variety who disapproved of it, none who had a purely old-fashioned education who were satisfied with it. The scientific men who came from the large public schools usually did nothing when there; they could not assimilate the subjects taught, and have abused the old system heartily. There are several serious complaints about superficial and bad teaching which I need not quote. Overteaching is thoroughly objected to; thus, in speaking of merits of education, I find:— 1. "Freedom to follow my own inclinations, and to choose my own subjects of study, or the reverse." 2. "The great proportion of time left free to do as I liked, unwatched and uncontrolled." 3. "Unusual degree of freedom."

I should add, that there are many touching evidences of the strong effect of home encouragement and teaching. As regards the subjects specially asked for, even by biologists, mathematics take a prominent place. Two of my correspondents speak strongly of the advantages derived from logic, and the weighty judgment of the late John S. Mill powerfully corroborates their opinions. Accuracy of delineation is also spoken of, and, owing to the extraordinary prevalence of mechanical aptitudes, I believe that the teaching of mechanical manipulation would be greatly prized. The interpretation that I put on the answers as a whole, is as follows: To teach a few congenial and useful things very thoroughly, to encourage curiosity concerning as wide a range of subjects as possible, and not to overteach.

As regards the precise subjects for rigorous instruction, the following seem to me in strict accordance with what would have best pleased those of the scientific men who have sent me returns:— 1. Mathematics, pushed as far as the capacity of the learner admits, and its processes utilized as far as possible for interesting ends and practical application. 2. Logic (on the grounds already stated, but on those only). 3. Observation, theory, and experiment, in at least one branch of science; some boys taking one branch and some another, to ensure variety of interests in the school. 4. Accurate drawing of objects connected with the branch of science pursued. 5. Mechanical manipulation, for the reasons already given, and also because mechanical skill is occasionally of great use to nearly all scientific men in their investigations.

These five subjects should be rigorously taught. They are anything but an excessive programme, and there would remain plenty of time for that variety of work which is so highly prized, as: ready access to books; much reading of interesting literature, history and poetry; languages learnt, probably best during the vacation, in the easiest and swiftest manner, with the sole object of enabling the learners to read ordinary books in them. This seems sufficient, because my returns show that men of science are not made by much teaching, but rather by awakening their interests, encouraging their pursuits when at home, and leaving them to teach themselves continuously throughout life. Much teaching fills a youth with knowledge, but tends prematurely to satiate his appetite for more. I am surprised at the mediocre degrees which the leading scientific men, who were at the Universities, have usually taken, always excepting the mathematicians. They prefer to fix of their own accord on certain subjects, and seem averse to learn what is put before them as a task. Their independence of spirit and coldness of disposition are not conducive to success in competition, they doggedly go their own way and refuse to run races.

Science has hitherto been at a disadvantage compared with other competing pursuits, in enlisting the attention of the best intellects of the nation, for reasons that are partly inherent and partly artificial. To these I will briefly refer in conclusion, with especial reference to the very important question, as to how far the progress of events tends to counterbalance or remove them.

If we class energy, intellect, and the like, under the general name of ability, it follows that, other circumstances being the same, those able men who have vigour to spare for extra professional pursuits, will be mainly governed in the choice of them by the instinctive tastes of their manhood. The majority will address themselves to topics nearly connected with human interests, a few only will turn to science. This tendency to abandon the colder attractions of science for those of political and social life, must always be powerfully reinforced by the very general inclination of women to exert their influence in the latter direction. Again, those who select some branch of science as a profession, must do so in spite of the fact that it is more unremunerative than any other pursuit. A great and salutary change has undoubtedly come over the feeling of the nation since the time when the present leading men of science were boys, for the state of education was then such as an enemy might have invented on purpose to exterminate science. It crushed the inquiring spirit, the love of observation, the pursuit of inductive studies, the habit of independent thought, and it protected classics and mathematics by giving them the monopoly of all prizes for intellectual work, such as scholarships, fellowships, church livings, canonries, bishoprics, and the rest. This gigantic monopoly is yielding, but obstinately and slowly, and it is unlikely that the friends of science will be able, for many years to come, to relax their efforts in educational reform.

As regards the future provision for successful followers of science, it is to be hoped that, in addition to the many new openings in industrial pursuits, the gradual but sure development of sanitary administration and statistical inquiry may in time afford the needed profession. These may, as I sincerely hope they will even in our days, give rise to the establishment of a sort of scientific priesthood throughout the kingdom, whose high duties would have reference to the health and well-being of the nation in its broadest sense, and whose emoluments and social position would be made commensurate with the importance and variety of their functions.

Chapter 14

Note and Queries on Anthropology, 1874: Statistics[1]

[Appendix] No. C—STATISTICS
By FRANCIS GALTON, Esq., F.R.S.

The topics suitable to statistics are too numerous to specify; they include every thing to which such phrases as "usually," "seldom," "very often," and the like are applicable, which vex the intelligent reader by their vagueness and make him impatient at the absence of more precise data. The principal things to be borne in mind in making statistical inquiries are:—

1. That the groups with which they deal should be homogeneous. *Ex.* It would be correct to inquire into the average height, and the frequency of different degrees of deviation from it, of greyhounds, and similarly as regards the speed of racehorses, but it would be absurd to talk of these averages as regards dogs or horses generally, because there are many varieties of them differing greatly and irregularly in height and speed (see further remarks in par. 5).

2. When the homogeneous groups are largely governed by the variation of a dominant influence, it is necessary to split them up into subdivisions, each referring to a short phase of the variation. *Ex.* It is correct to seek the average height of boys between 11 and 12, 12 and 13, and so on, but it would be absurd to seek that of boys generally.

3. To select cases on a system wholly independent of the quality about which the inquiry is made. *Ex.* It would be

[1] Galton 1874a.

correct to estimate the stature of the male adults of a nation by measuring individuals selected by lot; but it would be incorrect to take townsmen alone, and still worse those who lived in an unhealthy suburb of a town. It requires keen observation and much wariness to avoid errors due to a neglect of this caution, because phenomena that appear independent are often linked together in indirect and subtle ways. *Ex.* If we were to select the upper classes of Londoners according to the initial letter of their names, as printed in the Court Guide, and were to choose the letter Z for the purpose, we should be led utterly astray, as nearly all those names are foreign.

4. As regards the requisite number of cases, a few that are wisely selected and accurately reported are better than very many that are not. Each error sacrifices several good observations before it is diluted until it disappears; but if a faulty bias (as in par. 3) runs through all the observations, no increase in their number will eliminate it. Otherwise, the rule is that the precision varies as the square root of the number of observations; thus, twice the precision necessitates four times the labour. It is the best plan to proceed tentatively; if the results fall into more harmonious sequence as you proceed, it is worth proceeding; and if after dividing your statistics into 2, 3, or 4 groups you find the groups agree pretty well and that their sums form a yet more regular curve than that obtained from any of the subdivisions, you may safely trust it.

5. The law of deviation. Mention is above made of "homogeneous" groups: this epithet is applicable when individual differences are entirely due to the *aggregate* effect of a great many small and independent variable influences. *Ex.* The stature of an English male adult is due to his being a man of English race, reared under the range of those conditions of food, temperature, clothing, disease, and the like which prevail in England. The large causes common to all are the English breed and the range of English conditions; the small causes are differences of varieties and families, and of food, temperature, clothing, and the rest, within the range. The law of deviation depends wholly on the fact of *multifariousness of origin*; it has no more to do with the particular items of that multifariousness than the rules of arithmetic have to do with the quality of the things to be added or multiplied. Two and three make five, whether the objects be pence, or peas, or bills before Parliament; so the law of deviation holds for the stature of men and animals, and apparently, in a useful

degree, for every homogeneous group of qualities or compound qualities, mental or bodily, that can be named. It is a very general statistical law. The obvious effect of multifariousness is to make it an extremely rare event that all or nearly all the influences should be exerted in the same direction. *Ex.* It is a very rare event that all the cards in a hand at whist are found to be of the same colour. This is a simple result of the law of permutation: there are a vast and calculable number of different events each of which is equally likely to occur, and only one of these is the event in question. Proceeding on this principle and making certain rather forced suppositions to render calculation feasible, the law of deviation is mathematically deduced; and comparing fact with theory, wherever comparison is possible, it is found that they agree very fairly and in many cases surprisingly well. Reasoning backwards, we may suspect that a group is not homogeneous, or that the large influences are not sufficiently subdivided into phases (refer back to 3), when it does not conform to this law. The law shows that the frequency of small deviations must be very much greater than that of large ones, and that the larger the deviation is, whether above or below the average, so the frequency of the occurrence diminishes in an accelerating degree. It also shows, owing to the suppositions introduced, that the deviations on either side of the average are symmetrical; this is rarely strictly the case in nature. The rate of diminution, according to the above-mentioned theory, is shown in the following diagram,[2] in which I suppose 1000 men to be ranged in a long line in order of stature, beginning with the tallest at A and ending with the shortest at B. Then the middle man at C (500th in the scale) will be of the *average* height, and the 250th man at D will be as much taller than the one at C (owing to the symmetry of the curve) as the 750th, at E, is shorter. Knowing these two facts, the height at C (call it c) and the difference (call it r) between C and D (or C and E), we can tell the distribution of height all along the line until we come near the ends, where the run of the figures always becomes irregular. Thus it will be found that the height of the 90th man is $c+2r$ that of the 20th man is $c+3r$. Similarly the 910th man is $c-2r$, and the 980th is $c-3r$. The curved line of the diagram remains unaltered whatever may be the number of equal parts into which the horizontal line A B is supposed to be divided. If it referred to 100 men instead of 1000, the ordinates would

2 See Figure 14.1.

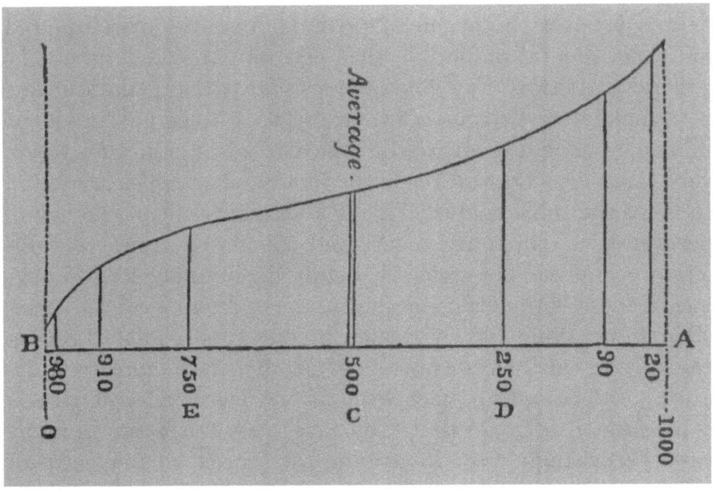

Figure 14.1: Inverted cumulative Normal distribution of stature, which Galton would later call the 'ogive'.

correspond to the 2nd, 9th, 25th, 50th, 75th, 91st, and 98th men instead of the 20th, 90th, &c.; consequently the traveller need only deal with a moderate number of men, arranging them in a series, and noting the height of those who stand at the *quarter points*. Thence he obtains c and r. He may do this for any quality whatever, running, shooting, colour, &c.; but it would be very proper that he should take notes concerning the men standing at the other points I have named, besides C and D. The common but much more laborious way of obtaining these results is to measure every individual and to group them into classes, so many per cent, between 5 feet 1 inch and 5 feet 2 inches, so many between 5 feet 2 inches and 5 feet 3 inches, and so on, from which data c and r can be obtained by the ordinary tables. Always mention the number of people measured, for the reliability of the average depends upon it; it is, as already mentioned, as its square root. A good book on these matters has yet to be written. Quetelet's *Letters on the Theory of Probabilities* is perhaps the most suitable to the non-mathematical reader; it is published in French in Brussels and is translated into English also (1849, Layton).[3]

[3] Quetelet 1849.

Chapter 15

The Spectator on Nuts and Men.

MR. GALTON ON NUTS AND MEN.[1]

MR. GALTON contributed a paper to the proceedings of the Royal Institution two or three months ago, which has just been published among the proceedings of the Institution, on the Nature and Nurture of Men of Science," and supplemented it by a remarkable discussion on the science of classification as illustrated by nuts. In one sense, this paper can hardly be called very instructive, its few facts of interest being a little jejune; but in another sense, it is really instructive, as an illustration of the singular weakness of the physico-scientific method when intruded on subjects to which it is not applicable. The paper is founded on the answers given to "a long and varied series of questions" asked by Mr. Galton of 180 Fellows of the Royal Society, of whom about half the number (80 or 90) have replied fully and minutely. Mr. Galton has elicited, first, that at least three-fourths of these distinguished men are possessed of a very eminent amount of vital energy. He gives a good many striking instances of this, but we do not dwell upon it, because, as he himself remarks, it is obvious that this is a trait not of men of science as such, but of remarkable men of any order whatever. Mr. Galton has occasion to distinguish sharply between the qualities of men of science and of commanders, statesmen, literary men," &c. Now, of course, energy as such belongs no more to one of these classes than to the others. Mr. Gladstone, who occupies his spare time in cutting down trees; Mr. Trollope, who for half the year is almost always hunting when he is not

[1] *The Spectator* May 16, 1874, 623–624.

adding to the productions of a talent more fertile of works than that of any previous literary man of this or any other age; Mr. Browning, who rejoices in feats of swimming almost as much as in feats of imaginative analysis of human character, are, to take one or two prominent examples, not likely to be deficient in vitality or vis. No doubt, in insisting on the astounding power of work and concentration which the leading men of science can command, Mr. Galton has hit upon one of the most marked criteria of intellectual distinction of any kind, not on one of those in any degree peculiar to men of science. Then, again, in like manner, for his second characteristic, that of good health and of descent from healthy parents, Mr. Galton does not claim any special appropriateness to men of science. It has been as marked a characteristic of eminent statesmen who have lived to the ages of Mr. Galton's correspondents, as of any other class of eminent men, except, perhaps, of poets of the less simple and more artificial kind. And so, again, of the gift of perseverance and of business habits, by which men of science are distinguished. Coming to the more special qualifications of men of science, as such, Mr. Galton assigns a habit of independence of mind, personal and inherited, as one of the most striking. He finds that they or their parents have very frequently belonged to a great variety of strange, small, or unfashionable sects; that many of them have had fathers of sturdy, not to say eccentric habits of mind,—that a large proportion had, or inherited, the habit of doing what they judged best, without deferring to social conventions. Then Mr. Galton finds a strong innate taste for some special branch of science to have been possessed by a large number of his correspondents. Again, most of their minds are directed to facts and abstract theories, and not to persons or human interests. The man of science is deficient in the purely emotional element and in the desire to influence the beliefs of others. Thus I find that two out of every ten do not care for politics at all; they are devoid of partisanship. They school a naturally equable and independent mind to a still more complete subordination to their judgment. In many respects their character is strongly anti-feminine. It is a curious proof of this that in the very numerous answers which have reference to parental influence, that of the father is quoted three times as often as that of the mother. It would not have been the case, judging from inquiries I elsewhere made, if I had been discussing literary men, commanders, or statesmen, or still more, divines."

If we add to this the discontent of men of science with the old classical system of education, and their wish for more various and stimulating school studies, we have in brief the substance of what Mr. Galton has distilled for us from the replies received to his queries. And we do not think that even our most careful readers will find it very instructive. For, in the first place, those of Mr. Galton's facts which look, on the surface of them, most interesting are quite too vague for any really instructive interpretation. Where, for instance, on Mr. Galton's principle, would he put the late Mr. John Stuart Mill? Among the men of science or the men of emotional human interests? He was a good botanist by innate taste. His interest even in men and minds was apt to be the interest of men of science in facts and abstract theories, "rather than in their human interests." Mr. Mill quotes his father habitually throughout his "Autobiography," and never mentions his mother once, except to condemn his father for marrying with inadequate means. And yet he makes a great deal more of his wife than of his father, exaggerates (in all probability) her true influence over his intellect to something like an hysterical extent, cares for politics with a passion far exceeding the passion of statesmen, and weaves the emotional element deep into his theory of society, of wealth, and of government. Again, where are you to rank the late Mr. Buckle, who referred to his mother with the kind of enthusiasm which Mr. Galton thinks scientific men are apt to show for their fathers, and yet who made the evolution of the purely intellectual principle the sole governing element in the history of civilisation, and devoted an almost womanish enthusiasm to the effort to demonstrate that faith and feeling have no important influence in determining the course of the world's progress? The truth is, that Mr. Galton is thinking solely of physical science, when he talks as he does about the character of the scientific man; and all that he proves is, that when you get a man of first-rate energy and power who is interested in the order of nature, but does not care much for mankind, his energy and power will be apt to produce results in the sphere to which it is confined. There is no great mystery in that. It is hardly a new discovery, that if you shut off a great force from the channels in which it is ordinarily frittered away, you will have a good deal more left for the particular mill-wheel on to which you direct it. True, or apparently true, abstract theories of things, are, as a rule, easier to reach than true or apparently true abstract theories of men; and hence, other things being equal, the minds of men devoted to physical science are apt to make more way than those devoted to the moral sciences. But nothing seems to be less probable than that mental, or moral, or political science can be cultivated with any real success without human interests, and without the emotional element." That students of Light, or Heat, or even Biology, should be all the more successful for not having had their interests distracted by excursions into the field of modern politics or literature is hardly remarkable. With a finite amount of vital energy to dispose of, every subtraction, however minute, will leave lees behind. Mr. Galton's facts, therefore, being founded on a narrow and one-sided notion of science, lose a good deal of their interest on anything like scrutiny. They come to very little more than this,—that discoverers must be original, and that investigators of physical truths will be all the more successful for not having too many and too intense moral interests. Take something appreciable from a given intellectual quantity, and something less will remain behind.

But Mr. Galton is more instructive in relation to the warnings he suggests, than in relation to the lessons he supplies. He is eloquent as to the principle of classifying the capacities of literary men, whom he proposes to deal with on the same Statistical Scale of averages by the help of which he might designate the size of nuts:—

"The law of statistical constancy may be taken for granted. It is evidenced by the experience of insurance offices against fire, death, shipwreck, and other contingencies, always with the proviso that the facts are gathered with discretion, on well-known general principles. Hence we may say with assurance, that although two common nuts may differ, yet the contents of different packets, each containing 1,000 nuts, will be scarcely distinguishable, for the same number of nuts of different sizes will be found in each. Let the contents of the several packets be each arranged in a long row, in order of size, beginning with the biggest nut and ending with the smallest, and place the rows rank behind rank; then by the law of statistical constancy the nuts in the same files will in all cases be closely alike (except the outside ones, where more irregularity prevails). Again, if we incorporate two rows into one of double length, still preserving the arrangement as to regular gradation in size, the centre nuts of the two original series will still be found at or near the centre of the compound series, the nuts in quarter positions will still be in quarter positions, and so on. Hence, whatever be the length of the series, the relative position in it of the nut will be a strict criterion of its size. This is of course equally true of all groups of qualities or characters whatever, in which the law of statistical constancy prevails, the series, in each case, being arranged according to gradations of the quality in question. Each individual is measured against his neighbour, and it is quite unnecessary to have recourse to any external standard. As regards a scale of equal parts, I make use of a converse application of the law of 'frequency of error' [this was illustrated by many experiments], which shows that in a row (say as before) of nuts, if we take those which occupy the three quarterly divisions (1st quarter, centre, 3rd quarter) as three elementary graduations of size, a continuous scale of graduations will be determined by the following series, in which the places of the nuts are supposed to be reckoned from the end of the row where the large nuts are situated, and to be given in per-thousandths of the entire length of the row. It might be called the Common Statistical Scale ' (S.S.).........The habit should therefore be encouraged in biographies, of ranking a man among his contemporaries, in respect to every quality that is discussed, and to give ample data in justification of the rank assigned to him. By the general use of a system like the above, which is universally applicable, social and political science would be greatly raised in precision."

Is it possible to conceive a quainter perversion of a scientific method, really applicable to magnitudes, to the purpose of a description of moral qualities? It is not difficult by eye or by more artificial measurements to arrange a quart of nuts so as to tell, by its place in the scale, the magnitude (approximately) of a particular nut. But the task of arranging a thousand men in relation (say) to their independence of character, or the strength of their "desire to influence the beliefs of others," or the magnitude of their political interests, or even their relative respect for their fathers' and mothers' influence over them, would somewhat overtask, we fear, all the resources of Mr. Galton. Nor do we see how his advice could be followed, unless the biographer of each man arranged the subject of his biography not in one such Statistical Scale, but in a few scores of them at least. What a delightful field to expatiate in, a biography written on such a principle would be! The late Dr. Arnold, for example, might, perhaps, be accepted as about the biggest nut of our time in the Statistical Scale, as regards "wish to influence the beliefs of others." But just conceive the usefulness of a discussion as to whether Mr. Matthew Arnold equals, or exceeds, or falls short of, his father in this quality. Imagine a keen analysis of "Literature and Dogma" or of "The Garland of Friendship" hung on the tablet to the memory of "Arminius," or of his poems, or his book on the German Universities, made with a view to solve this important problem; conceive the utility of making a careful comparison between these considerable men and Mr. Gladstone, Bishop Wilberforce, and Lord Salisbury, with a view to determine where they stand between 4° S.S. and —4° S.S. in relation to this same quality, and then consider what a very small proportion of the work to be done would have been done. If Mr. Galton's wonderful principle is to be followed, the scale must be made up not in relation to this quality only, but to most other distinctive qualities. You must say whether Dr. Arnold was 2° S.S. in relation to historical acumen, or —3° S.S., or where. Then you must deal with him as teacher, and place him among the large or small nuts in relation to the power to impart knowledge; then, again, as preacher, and estimate the point to which he rises in the Statistical Scale as a successful imparter of moral influence. Again, you must deal with his political sagacity in the same fashion, and determine whether he is a big or a small nut in that; and when you have done all this, you will have wasted your labour, for you will not have distinguished between a desire to influence beliefs arising from one source, such as earnest conviction, and a desire arising from another source, such as ambition, or between political acumen due to historical enthusiasm, and political acumen due to intimate knowledge of the people. In a word, Mr. Galton's attempt to give a scientific turn to biography would spoil literature, without enriching science.

And that is why we say that Mr. Galton's paper is instructive. It shows how helplessly men with the physico-scientific bias upon them flounder about in the region of human and moral influences, if they have not taken care to balance their judgment by the cultivation of the Literae Humaniores. Let them prefer their fathers' opinions as earnestly as they please to their mothers'; let them despise the old classical discipline with all their hearts; let them have had a taste for bones or stones from their childhood upwards; let them boast of the most perfect of all combinations between active brains and weight of character; let them have been educated from infancy to let the opinion of the world run off them as water runs off a duck's back, whenever they think it unreasonable; but yet let them not attempt to help the world in the estimate of human character, by suggestions derived chiefly from the world of physical science. If they do, they will come to grief, as Mr. Galton has come to grief, by suggesting methods which plunge literary men into wonder and awe at the clumsiness, uselessness, inefficacy, and complete inapplicability to the subject of the tests and gauges suggested. A distinguished surgeon's proposed prayer-gauge produced a great bewilderment and amusement the other day in the minds of all really devotional people. But it was not nearly so ludicrous as Mr. Galton's suggestion for making a Statistical Scale of mental and moral qualities, on the same principle on which he obtains one whereby you can assign the magnitude of any given nut. What he has first to do is to suggest any test of relative magnitude as to matters so complex and delicate as moral qualities, especially where differences of kind invariably accompany and profoundly affect differences of amount. Mr. Galton might almost as well seriously apply the theory of chemical affinities to moral character, and try to assign chemical equivalents to the qualities of generosity, magnanimity, ambition, and so forth, as recommend us to range a thousand men in the relative order of their share in a given quality, just as you range the nuts out of a quart pot, and so obtain a scale of mental and moral magnitude. Mr. Galton's "Statistical Scale" will never take any hold on the world, because it is wholly and intrinsically inapplicable to the purposes for which he recommends it. But if ever it did attain an unfortunate notoriety, we suspect it would be much more likely to be called the scale of Sham Science, than by the name by which its inventor has proposed to christen it.

Chapter 16

A Theory of Heredity

Mr. Darwin stated,[1] in the year 1868, in the preface to his theory of Pangenesis,[2] that "every one appears to admit that the body consists of a multitude of 'organic units,' each of which possesses its own proper attributes, and is to a certain extent independent of all others;" and it may be safely asserted that the general expression of biological opinion since that date has been emphatically the same. We may therefore rest assured that the hypothesis of organic units, and all that such an hypothesis implies, must lie at the foundation of the science of heredity. It remains to determine further particulars; we have to examine how far the details of such theories as are based upon the hypothesis of organic units are correct, and to consider how their deficiencies may be supplied.

The facts for which a complete theory of heredity must account may conveniently be divided into two groups; the one refers to those inborn or congenital peculiarities that were also congenital in one or more ancestors, the other to those that were not congenital in the ancestors, but were acquired for the first time by one or more of them during their lifetime, owing to some change in the conditions of their life.

The first of these two groups is of predominant importance, in respect to the number of well-ascertained facts that it contains, many of which it is possible to explain, in a broad and general way, by more than one theory based on the hypothesis

[1] Galton: This memoir was in type for publication in the December number of the *Contemporary Review*, at the time when it was read before the Anthropological Institute. I have taken the opportunity afforded by a reprint, to revise it considerably, and to amend it in many particulars.—F.G.

[2] Galton: Darwin: "Variation of Plants and Animals under Domestication," ii. 370.

of organic units. The second group includes much of which the evidence is questionable or difficult of verification, and which, as I shall endeavour to show, does not, for the most part, justify the conclusion commonly derived from it. In this memoir I divide the general theory of heredity into two parts, corresponding respectively to these two groups. The first stands by itself, the second is supplementary and subordinate to it.

No theory of heredity has been enunciated with more clearness and fulness than that of Mr. Darwin's Pangenesis, and the preparatory statement to that theory contains the most elaborate epitome that exists, of the many varieties of facts for which a complete theory of heredity must account. What I have now to say, is largely based on the arguments and considerations brought forward by Mr. Darwin in support of Pangenesis; nevertheless the conclusions in this paper will be seen to differ essentially from his own. Pangenesis appears more especially framed to account for the cases which fall in the second of the above-mentioned groups, which are of a less striking and assured character than those in the first group, and it will be seen that I accept the theory of Pangenesis with considerable modification, as a supplementary and subordinate part of a complete theory of heredity, but by no means for the primary and more important part.

Before proceeding, I beg permission to use, in a special sense, the short word "stirp," derived from the Latin *stirpes*, a root, to express the sum-total of the germs, gemmules, or whatever they may be called, which are to be found, according to every theory of organic units, in the newly fertilised ovum—that is, in the earliest pre-embryonic stage—from which time it receives nothing further from its parents, not even from its mother, than mere nutriment. (It is hardly necessary to remind the reader that not a drop of blood from the mother penetrates into the vessels of the embryo, but that the two circulations are wholly distinct; the placenta to which the embryo is attached, and with which it is in vascular connection, being itself nourished from the mother by mere imbibition.) This word "stirp," which I shall venture to use, is equally applicable to the contents of buds, and will, I think, be found very convenient, and cannot apparently lead to misapprehension.

The whole of the "stirp," together with much of mere nutriment, is packed into a space not exceeding the size of the head of a pin, for that is about the size of the newly fertilised ovum, which, curiously enough, is the same in all mammalia. It is evident that direct observation can tell us nothing concerning the form and behaviour of such minute objects as the germs of which the stirp is composed; they would be far beyond the ken of the microscopist, even if their separate actions upon light were different. But even this is not the case, for the fertilised ovum is almost homogeneous in colour. Ova and their contents are, to biologists looking at them through microscopes, much what mail-bags and the heaps of letters poured out of them are to those who gaze through the glass windows of a post office. Such persons may draw various valuable conclusions as to the postal communications generally, but they cannot read a single word of what the letters contain. All that we may learn concerning the constituents of the stirp must be through inference, and not by direct observation; we are therefore forced to theorise.

We will begin with a statement of the four postulates that seem to be almost necessarily implied by any hypothesis of organic units, and which are included in that of Pangenesis. The first is, that each of the enormous number of quasi-independent units of which the body consists, has a separate origin, or germ. The second is, that the stirp contains a host of germs, much greater in number and variety than the organic units of the bodily structure that is about to be derived from them; so that comparatively few individuals out of the host of germs, achieve development. Thirdly, that the undeveloped germs retain their vitality: that they propagate themselves while still in a latent state, and contribute to form the stirps of the offspring. Fourthly, that organisation wholly depends on the mutual affinities and repulsions of the separate germs; first in their earliest stirpal stage, and subsequently during all the processes of their development.

Proofs of the reasonableness of these postulates are especially to be found in the arguments of Mr. Darwin: that there is at least a fair ground to believe in their reasonableness, may be shown in a cursory manner. Thus, the independent origin of the several parts of the body may be argued from the separate inheritance of their peculiarities. If a child has its father's eyes and its mother's mouth, these two features must have had a separate origin. Now, it is observed that peculiarities, even of a microscopic kind, are transmissible by inheritance, and therefore it may be concluded that the minutest parts of the body have separate origins. That the stirp contains a much greater variety of germs than achieve development is proved by the fact that a person is capable of transmitting a variety of ancestral peculiarities to his children, that he did not himself possess. But since everything that reached him from his ancestors must have been packed in his own stirp, it follows that his stirp contained in addition to such peculiarities as were developed in his own bodily structure, those numerous other ancestral peculiarities of which he was personally destitute, but which he bequeathed to one or more of his descendants. Therefore every stirp must be held to contain a great variety of germs in addition to those that may achieve development in the person who grows out of that stirp. It further follows, that these residual germs retain their vitality, and contribute to form the stirp of the descendants, as will be explained more fully further on. The fourth and last postulate, that organisation wholly depends on the mutual affinities of the separate organic units, commends itself to acceptation by the simplicity and sufficiency of what is asked; much of what I have to say in this paper, will testify to this. We must also bear in mind, that the alternative hypothesis of a general plastic force resembles that of other mystic conceptions current in the early stages of many branches of physical science, all of which yielded to molecular views, as knowledge increased. The science of heredity is still in an early stage, and analogy disposes us to expect that its course will be similar to that of its predecessors. The possibility of such minute objects as the germs possessing sufficient delicacy of perception to ensure that each of so enormous a variety of them should find its place, was illustrated by Mr. Darwin through the delicate perception of the pollen grains of the different species of plants. He says:[3] "About 10,000 compositae exist, and there can be no doubt that if the pollen of all the species could be simultaneously or successively placed on the stigma of any one species, this one would elect, with unerring certainty, its own pollen." The partial failures in the action of these affinities are most instructive, as where a mark of any kind on the skin is transmitted by inheritance in an altered situation, to a neighbouring or to an homologous part. Having stated thus much by way of preface, we may now proceed freely.

Much wonder is expressed by physiologists at the apparent fact that none, at least of the higher races, admits of being

---

[3] Galton: "Variation of Plants and Animals under Domestication," ii. 380.

long maintained through any system of unisexual parentage; but that a deterioration, which we may reasonably ascribe to a deficiency of some of the structural elements, is always observed to set in and gradually to increase, the race ultimately perishing from that cause. A system of double parentage is therefore a very important requirement, some think an essential one, to secure the indefinite maintenance of any race whose organisation is complex. What is the explanation of this? In the first place, double parentage should be looked upon as the primary requirement, and sex as the consequence, not the cause of that requirement. There are not of a necessity two sexes, because swarms of creatures of the simplest organisations mainly multiply by some process of self-division. On the other hand, as I shall endeavour to show, there is a theoretical advantage in a system of double parentage, which grows to be of paramount importance, as organisations increase in complexity. But it is through the evolution of sex, that a system of double parentage is secured, and, therefore, I would submit, it is to the need of the latter that we must ascribe the existence of the former. The opposite view is certainly erroneous, namely, that sex is an inherent necessity, and that double parentage is the simple consequence of it. Such a sequence, which, perhaps, represents the common and unreasoned theory, is, to use a common phrase, "putting the cart before the horse." As has been just remarked, in many of the lowest forms of organised life, double parentage exists, but sex apparently does not, because any two cells seem able to conjugate and to combine their contents within a single cell; these forms are also capable of easy unisexual multiplication by self-division or by budding. Proceeding higher up the scale of life, the sexual differentiation becomes increasingly marked, and unisexual propagation is of rarer occurrence. At length we reach a stage where the differentiation of sex is complete, and the power of unisexual propagation is wholly lost. Now the necessity of a system of double parentage in complex organisations, is the immediate consequence of a theory of organic units and germs, as we shall see if we fix our attention upon any one definite series of unisexual descents, and follow out its history. Suppose we select, cut off, and plant the second bud, then after it has grown to maturity we similarly take the second of *its* buds, and so on consecutively. At each successive stage there is always a chance of some one or more of the various species of germs in the stirp dying out, or being omitted;

and of course when they are gone they are lost for ever, and are irreplaceable by others. From time to time this chance must fall unfavourably, and will cause a deficiency in some of the structural elements, and a consequent deterioration of the race. If the loss be vital, this particular line of descent will of course be extinguished at once; but on the more favourable supposition, the race will linger on, submitting to successive decrements in its constituent elements, until the accumulation of small losses becomes fatal. What is true for the series of second buds in our example, is of course equally true for any system we please to specify, and therefore it would be generally true in the experience of gardeners and others.[4]

But in a free state of nature, where the weakly plants are supplanted by those that remain sound, a new consideration is introduced. Here we have to consider, on the one hand, the growing chance against the deterioration of each single line of descent, and on the other, the growing number of all possible lines of descent. They both proceed in a geometrical ratio; and if the ratio of the latter exceeded that of the former, extinction need not take place. But, again, this excess would become an impossibility after a certain degree of complexity had been reached, because with growing complexity, the chance of deterioration must increase, while the fecundity (see H. Spencer's 'Biology,' vol. i. "Multiplication") necessarily diminishes. On the other hand, when there are two parents, and therefore a double supply of material, the chance deficiency in the contribution from either of them, of any particular species of germ, tends to be supplied by the other. No doubt, cases must still occur, though much more rarely than before, in which the same species of germ is absent from the contribution of both, and a very small proportion of the families will thereby perish. But what if they do become extinct? The remaining families are perfectly sound, or tend to become so in each succeeding generation, and they fill up, only too easily, the gap. Thus we see that in any specified course of unisexual generation, every line of descent is doomed to extinction, sooner or later; but that in bisexual, only a very small proportion of families become extinct, or even temporarily suffer, from the cause we are considering, while the great majority do not suffer a whit,

---

[4] Galton: It might be worth the while of the mathematical reader to refer to a paper on an analogous subject, "The Extinction of Surnames," by the Rev. H. W. Watson, in the Journal of the Anthropological Institute, 1874, p. 138, to which there is a page of preface by myself.

and those few who do, tend to become rehabilitated. There is yet another advantage in double parentage, namely, that as the stirp whence the child sprang, can be only half the size of the combined stirps of his two parents, it follows that one half of his possible heritage must have been suppressed. This implies a sharp struggle for place among the competing germs, and the success, as we may infer, of the fitter half of their numerous varieties.

The limitation of space in the stirp must compel a limitation not only to the number of varieties of each species of germ, but also to the number of individuals in each variety. The knowledge of such a fact is helpful, and appears to be needed, in accounting for the not very large number of subdivisions in which peculiarities are transmitted. I am not now considering cases of the slow loss of some characteristic of a race, which proceeds by minute gradations, and which may be ascribed, at least in part, to a change in the quality of the germs, nor am I now speaking of cases where it is clear that one of two alternative qualities has overpowered the other, but of instances where they are equipotent and in no way antipathetic. Thus, in the gradual breeding-out of negro blood, we may find the colour of a mulatto to be the half, and that of a quadroon to be the quarter of that of his black ancestors; but as we proceed further, the subdivision becomes very irregular; it does not continue indefinitely in the geometrical series of one-eighth, one-sixteenth, and so on, but it is usually present very obviously, or not at all, until it entirely disappears. There are many more gradations in compound results, as in an expression of the face, because any one of its elementary causes may be present or absent; and as the number of possible combinations or alternatives, among even a few elements, is very great, there must be room for a large number of grades between the complete inheritance of the expression and its total extinction.

It is certain, from the rapidity of the visible changes in the substance of the newly fertilised ovum, that the germs in the stirp are in eager and restless pursuit of new positions of organic equilibrium, due, as we may suppose, to the unequal rates of development of some of the better nourished germs. We see that segregations occur as much as aggregations, and it is reasonable to suppose that repulsions concur with affinities in producing them. We know nothing as yet of the nature of these repulsions and affinities, but it seems hardly possible to account for the whole state of affairs on the hypothesis of a purely step-by-step development like that proposed in Pangenesis, where B follows A, and C follows B, and so on. It is difficult to suppose the directions of the mutual influences of the germs to be limited to lines, like those that cause the blood- corpuscles to become attached face to face, in long rouleaux, when coagulation begins; neither can we suppose them limited to planes, like those that govern the harmonious groupings of the flora and fauna on the face of a land left in a state of nature; but we ought rather to expect them to act on many sides, in a space of three dimensions, just as the personal likings and dislikings of an individual in a flying swarm may be supposed to determine the position that he occupies in it. Each germ has many neighbours: a sphere surrounded by other spheres of equal sizes, like a cannon ball in the middle of a heap of them, when they are piled in the most compact form, is in immediate contact with no less than twelve others. We may therefore feel assured, that the germs must be affected by numerous forces on all sides, varying with their change of place, and that they must fall into many positions of temporary and transient equilibrium, and undergo a long period of restless un- settlement, before they severally attain the positions for which they are finally best suited. However ignorant we may be at present of the character of these affinities and repulsions, or of what Mr Herbert Spencer calls their polarities, in his instructive chapters in the first volume of his "Principles of Biology," a conviction of their existence is sufficient to afford general notions of what must be their mode of action, and enables us to illustrate its necessary consequences by many familiar experiences. Chief among these are the events of political life, such as those connected with the struggle for place and power, with election, and with representation. We know that the primary cells divide and subdivide, and we may justly compare each successive segmentation to the division of a political assemblage into parties, having, thenceforward, different attributes. We may compare the stirp to a nation and those among its germs that achieve development, to the foremost men of that nation who succeed in becoming its representatives; lastly, we may compare the characteristics of the person whose bodily structure consists of the developed germs, to those of the house of representatives of the nation. These are not idle metaphors, but strict analogies; they will be found to bear consideration, and to be worthy of being pursued, as they give a much-needed clearness to views on heredity.

The great dissimilarity frequently observed between brothers or sisters is to be accounted for and easily illustrated by a political metaphor. We have to recognise, on the one hand, that the stirps of the brothers and sisters must have been nearly alike, because the germs are simple organisms, and all such organisms breed true to their kind, and on the other hand, that very different structures have been developed out of those stirps. A strict analogy and explanation of all this is afforded by the well-known conditions and uncertainties of political elections. We have abundant experience that when a constituency is very varied, trifling circumstances are sufficient to change the balance of parties, and therefore, although there may be little real variation in the electoral body, the change in the character of its political choice at successive elections may be abrupt. A uniform constituency will always elect representatives of a uniform type; and this result precisely corresponds to what is found to occur in animals of pure breed, whose stirp contains only one or a very few varieties of each species of germ, and whose offspring always resemble their parents and one another. The more mongrel the breed, the greater is the variety of the offspring.

In twins of the same sex a dissimilarity is not unfrequently found of a more marked description than that between ordinary brothers and sisters, notwithstanding that the embryonic conditions of the twins must have been closely similar. This is a very curious subject, and requires the following explanation. I had occasion to make many inquiries into the resemblances of twins, whence it appeared that among well-formed "true" twins,[5] so to speak, namely, those who, up to the time of their birth, were enclosed in the same membrane, and had therefore been developed out of two germinal spots in the same ovum,

[5] For some general results of these inquiries, see the paper printed in the miscellanies of this volume [Galton 1875]. I had twenty cases of strong dissimilarity in twins, and in all the cases the twins were of the same sex. Now it appears to be a rule without exception that what I have above termed "true" twins are of the same sex. Such twins are by no means uncommon; Spaeth's estimate of their frequency, as compared to that of twin births generally, is as high as 25 per cent., and I understand that his observations rank among the very best; however, the estimates of other observers are much lower. Hence there is much probability that my cases of strong dissimilarity were usually, if not invariably, cases of true twins. But I have no direct evidence one way or the other.

there exist two groups of cases that contrast strangely with one another, while there are, comparatively speaking, only a few intermediate cases. In the larger of the two groups, the twins are exceedingly alike in body and mind; also in their growth, illnesses, and decay, and their resemblance is not unfrequently such as to justify the somewhat startling incidents referring to twins, that are to be found in many works of fiction. In the smaller group, which contains perhaps one-fourth as many cases as the larger, the twins are absolutely unlike; so much so, that they have occasionally been described as "complementary" the one to the other—the one having what the other lacked. What can be the reason that, out of identically the same primary stirp, either two absolutely dissimilar persons can be developed, or else two closely similar ones; while the intermediate cases are comparatively rare, so that they may be considered due to quite another and more common contingency—namely, that in which the twins are not produced out of the same ovum, but from separate ova? The answer I suggest is as follows:—As regards the similarity of true twins, there can be little difficulty; we should expect, on statistical grounds, that the two halves of any assemblage of germs would be much alike. The secondary stirps of the twins being alike, and the circumstances under which the bodily structure is developed out of them being almost identical, the results must be closely similar. On the other hand, as regards the dissimilarity, we might expect that if there had happened to be a sufficient delay before the division of the primary stirp, to allow its germs to arrange themselves somewhat according to their affinities, the twin halves of the primary stirp would be strongly contrasted. Political analogies may again be appealed to with advantage. In the case of an ordinary single birth, each germ that achieves development may be compared to the sole representative of a body of electors, each of whom has a single vote; then, in the case of twins, two representatives have to be elected, each elector still having only a single vote. If one of the political parties slightly predominates, and if the electoral body be divided by an accidental line, the same party would predominate in each division; consequently if the election happened to be so conducted, both representatives would be men of the same predominant party, and of identical politics. But if the electoral body acted as a whole, the predominating party would be unable to return more than a single candidate; consequently the two representatives would be men of opposite politics.

Individual variation depends upon two factors; the one, is the variability of the germ and of its progeny; the other, is that of all kinds of external circumstances, in determining which out of many competing germs, of nearly equal suitability, shall be the one that becomes developed. The variability of germs under changed conditions, and that of their progeny, may be small, but it is indubitable; absolute uniformity being scarcely conceivable in the condition and growth, and, therefore, in the reproduction of any organism. The law of heredity goes no further than to say, that like *tends* to produce like; the tendency may be very strong, but it cannot be absolute. The effect of the second of the factors mentioned above, is that a very slight variation in the germs may have a momentous effect in the personal structure that is developed out of a comparatively small number of them. There are numerous competing germs in the place of each unit of structure, and only one of the competitors can succeed in each case. When the competition is close, a very slight difference, either in the intrinsic quality of the germ, or in its temporary position at some critical time, or in any other variable circumstance, may determine its success. It may well happen, that some species of germs may have failed in achieving development during very many generations, by the end of which time they may have become considerably modified, and at length, partly owing to their intrinsic improvement, and partly to the accident of favourable circumstances, that species may suddenly vanquish its competitors and achieve development, and be the cause of a marked individual variation. The chain of sequences would have been perfectly continuous, though its manifestation in the form of personal structure would appear strangely capricious. Precisely similar catastrophes are of notoriously frequent occurrence in political and social life.

Those germs which have become developed into cells, have been supposed (I believe universally) to be the important, if not the chief agents in maintaining the progeny of germs; in other words, of influencing heredity. This is certainly an essential condition in the theory of Pangenesis, as the name of that theory testifies; there, each separate cell in its nascent state is supposed to throw off germs that circulate freely in the body along with the others that had been hereditarily transmitted. It is from among the general mass of these, that certain groups are supposed to aggregate themselves, owing to their mutual affinities, and so to form the sexual elements. For my own part, while acknowledging that there is undeniable evidence of the existence of the power of cells to throw off germs, which will be discussed when we come to the second group of cases, I shall endeavour to show that its effects on inheritance generally, are minute and secondary. My argument is this: Of the two groups of germs, the one consisting of those that succeed in becoming developed, and in forming the bodily structure, and the other consisting of those that remain continually latent, the latter vastly preponderates in number. We should expect the latent germs to exercise a corresponding predominance in matters of heredity, unless it can be shown that, on the whole, the germ that is developed into a cell, becomes thereby more fertile than if it had remained latent. But the evidence points the other way. It appears both that the period of fertility is shorter, and the fecundity even during that period is less in the germ that becomes developed into a cell, than they are in the germ that remains latent. Much less then would the entire bodily structure, which consists of a relatively small number of these comparatively sterile units, successfully compete in matters of heredity with the total effect of the much more numerous and more prolific units which are in a latent form. The shortness of the period of fertility of the germ that becomes developed, is implicitly acknowledged even by the author of Pangenesis, who considers it to cease so soon as the cell is completely formed (*op. cit.* ii. 374), and the hypothesis that the developed germ is less fecund, even during its short period of fertility, than the germ that continues latent, agrees singularly well with many classes of fact. Thus it explains why, although hereditary resemblance is the general rule, the offspring is frequently deficient in the very peculiarity for which the parent was exceptionally remarkable. We can easily understand that the dominant characters in the stirp will, on the whole, be faithfully represented in the structure of the person who is developed out of it; but if the personal structure be a faithful representative of the dominant germs, it must be an over-favourable representative of the germs generally, and therefore, *à fortiori*, of the undeveloped residue; nay, in extreme cases, the personal elements may be absolutely misrepresentative of the residual elements, the accidental richness of the sterile sample in some particular valuable variety of germ, having drained the fertile residue of every germ of that variety. The possibility of this occurrence is the more credible, since, as we have already seen, the number of germs of each variety cannot be very large. Experience testifies to the fact that children of men of extraordinary genius have not unfrequently been singularly deficient in ability, and this condition has been especially remarked in instances where the man of genius was himself the offspring of a mediocre ancestry; where, therefore, according to the above theory, the number of valuable germs were few, and all of them were used up and rendered comparatively sterile in the structure of his own person. The steady tendency to deterioration in exceptional peculiarities is likewise shown by the avowed difficulty, among breeders, of maintaining the high character of any valuable variety that has been produced by accident (that is, by some happy combination of a number of unknown variable causes). Another result of the best elements of the stirp being rendered sterile, is the strong tendency to deterioration in the transmission of every exceptionally gifted race. That this is a universal tendency among races in a state of nature, is proved by the fact that existing races are only kept at their present level by the severe action of selection. If they were left unpruned even for a single generation, the weaker members would survive, and the average quality of the race would necessarily diminish.

Again, the sterility of the developed elements of the stirp explains the fact of certain diseases skipping one or more generations, if the further very reasonable postulates are granted, that the germs of those diseases are both prolific and gregarious. Thus, nearly all the gout molecules in the stirp whence A sprang might, owing to their gregarious nature, become developed in the person of A, and so be rendered sterile; the small fertile residue in his stirp would be insufficient to supply that of his son B with enough gout germs to dominate and achieve development in the person of B, consequently they would be husbanded; then, owing to their prolific character, they would so multiply in a latent form in the structure of B, as to insure transmission in sufficient numbers to the stirp of G the son, or D the grandson, to enable them to achieve development in the person of C or D, just as they had done in that of A; and so the cycle would be repeated.

The conclusion from what has thus far been said is amply confirmed by observation; it is:—1. That the contents of the stirp must segregate themselves into divisions or septs, and that these septs must subdivide again and again, under the influence of the mutual attractions and repulsions of their units, just as a large political party may repeatedly subdivide itself into different factions. (2.) That the dominant germs in each successive sept are those that achieve development. (3.) That it is the residual germs that are the parents of the sexual elements or buds.

No process of subdivision like that which has just been described could be expected to take place with perfect accuracy; no political party was ever split with such clean precision into two political septs, that none of the A party were included in the ranks of B, and *vice versâ*. We must therefore feel assured that germs of many alien species would be included in each successive sept. Also, we may reasonably suppose that the structure formed out of those germs that have developed into cells, must afford many convenient places for the lodgment and sustenance of the alien germs; consequently, representatives of all parts of the residue of the stirp would be found dispersed all over the body. Lastly, we cannot but expect that these alien germs, when they thrive and multiply, would somewhat transgress the bounds of the cell or cell-interspace in which their progenitors had lodged, knowing that even so large an object as a blood-corpuscle will occasionally find its way through the unruptured wall of a capillary vessel. This is a very different supposition to that of the free circulation of gemmules in Pangenesis, yet it seems to have the merits of that theory (so far as the group of cases are concerned which we are now considering, namely, the inheritance of qualities that were congenital in the ancestry), and at the same time to be free from the many objections that are urged against the theory of Pangenesis. These are as follows:—On physical grounds, we cannot understand how colloid bodies, such as the Pangenetic gemmules must be, could pass freely through membranes. Moreover, if they did, the paternal gemmules in the body of the unborn child would diffuse themselves equally over the body of the child and that of its mother; consequently there would be very few remaining in the body of the child, while, on the other hand, there would be an invasion of maternal gemmules. The final result of this would be, that the individual would transmit his or her maternal peculiarities far more than his or her paternal ones; in other words, people would resemble their maternal grandmothers very much more than their other grandparents, which is not at all the case. That the gemmules are not contained, in any large number, in the blood-vessels, is proved by my own experiments, in which I largely transfused the blood of an alien species of rabbit into the blood-vessels of male and female silver-grey rabbits, from which I afterwards bred. I repeated this process for three generations, and found not the slightest sign of any deterioration in the purity of the silver-grey breed.[6]

Again, a free circulation of the gemmules, such as Pangenesis supposes, would cause various events to be extremely common, whereas the supposition of their transgression through a small space beyond their original limits, shows them to be possible, though infrequent, just as they actually are. I mean such cases as the zebra-marks on the foal out of a thoroughbred mare by a thoroughbred horse, owing to the former having once borne a mule to a zebra; the action of pollen on the tissues adjacent to the fertilised pistil of a different variety of plant. The distribution of the germs, by the agency I supposed, all over the body, would account equally with Pangenesis for the replacement of a lost limb in the lower animals, and the reparation of simple tissues in the higher ones. It would much transcend my limits if I were to enter at length into these and kindred questions; but it is not necessary to do so, for it is sufficient to refer to Mr. Darwin's work, as already quoted, where they are most fully and carefully discussed, and to consider, while reading it, whether the theory I have proposed could, as I think it might, be substituted with advantage for that of Pangenesis. I must repeat, that I limit these remarks to the very large pro- portion of cases that fall into the first of the two groups, in which I am discussing the facts of heredity.

It will be convenient at this place to contrast the views that have been thus far set forth with those of Mr. Darwin in his theory of Pangenesis. That theory affirms as follows:—

(1.) There are cells, and there are a great number of gemmules.

(2.) The cells multiply by self-division, and during this process they throw off gemmules. [I look upon this process of throwing off gemmules to be of such minor importance as to have no effect whatever upon the cases we have thus far

---

[6] Galton: The experiments on the first generation were published Proc. Royal Society, 1871, p. 393; but see Mr. Darwin's remarks in *Nature*, 1871, p. 502, as to my conclusions. I subsequently carried on the experiments with improved apparatus, and on an equally large scale, for two more generations. [Darwin 1871; Galton 1871a.]

considered. Its existence is granted, but only as a subordinate process, to account for the exceptional cases to be hereafter considered, and not as the primary process in heredity.]

(3.) The gemmules multiply by self-division, and any gemmule admits, under favourable circumstances, of being developed into a cell. [I look upon this as the primary process in heredity.]

(4.) The personal structure is formed by a process analogous to the fertilisation of each gemmule that becomes developed into a cell, by means of the partially developed cell that has preceded it in the regular order of growth. [I look on it as due, first, to the successive segmentations of the host of gemmules that are contained in the newly fertilised ovum, owing to their mutual affinities and repulsions; and, secondly, to the development of the dominant or representative gemmules in each segmentation, the others remaining dormant, and are called, for convenience, in the next paragraph, the "residue."

(5.) The sexual elements are formed by aggregations out of the gemmules, all of which are supposed to travel freely throughout the body. [I look on the sexual elements as directly descended from the "residue," and do not suppose the gemmules to travel freely. I allow some very moderate transgression across the bounds of their domiciles, and something more than that, under the limitations that will be described in the latter part of this memoir.]

[I account for all varieties of the gemmules being found in all parts of the body, by the above-mentioned segmentations being never clean and precise. Hence it follows that each segmentation must contain stray and alien gemmules, and I suppose that many of these become entangled and find lodgment in the tissue developed out of each segmentation.]

We will next proceed to examine the cases that fall into the second group; they are those in which characters created artificially in the person of the parents, are transmitted by inheritance to their offspring. In considering what appear at first sight to be cases in evidence of this, we must be extremely careful not to confuse the effects of totally different processes.

We have thus far dealt with three agents—(1) the stirp, which is an organised aggregate of a host of germs; (2) the personal structure, developed out of a small portion of those germs; and (3) the sexual elements, generated by the residuum of the stirp. The cases before us are those which are supposed to prove that 2 reacts on 3—that is, the personal structure upon the sexual elements. The first and the largest class of these cases refer to adaptivity of race. It is said that the structure of an animal changes when he is placed under changed conditions; that his offspring inherit some of his change; and that they vary still further on their own account, in the same direction, and so on through successive generations, until a notable change in the congenital characteristics of the race has been effected. Hence, it is concluded that a change in the personal structure has reacted on the sexual elements. For my part, I object to so general a conclusion, for the following reasons. It is universally admitted that the primary agents in the processes of growth, nutrition, and reproduction are the same, and that a true theory of heredity must so regard them. In other words, they are all due to the development of the same germinal matter, variously located. Consequently, when similar germinal matter is everywhere affected by the same conditions, we should expect that it would be everywhere affected in the same way. The particular kind of germ whence the hair sprang, that was induced to throw out a new variety in the cells nearest to the surface of the body under certain changed conditions of climate and food, might be expected to throw out a similar variety in the sexual elements at the same time. The changes in the germs would everywhere be collateral, although the moments when any of the changed germs happened to receive their development, might be different. So far from there being evidence that the changed structure of the hair causes the germs in the sexual organs to vary, it may often happen that the latter are the first to change. Thus the progeny of thick-fleeced sheep, newly imported into the tropics, may begin to lose wool earlier than their parents. There is not a shadow of proof that the adaptivity of a race to changed conditions, *affecting all parts of the body alike*, is due to the reaction of changed personal structure upon the sexual elements. Another instance of simultaneous action is to be found in the fact that a drunkard is often known to have imbecile children, although his offspring previous to his taking to drink were healthy. The alcohol pervades his tissues, and, of course, affects the germinal matter in the sexual elements as much as it does that in the cells which form the structure of his own nerves. Exactly the same result must occur in the case of many constitutional diseases that have been acquired by long-continued irregular habits. The case is different as regards those conditions that have a local influence; but races are very slow in adapting themselves to these.

Another class of evidence brought forward in proof of the inheritance of non-congenital peculiarities concerns mutilations. No doubt the industry of M. Prosper Lucas, and of many others, has brought together several curious cases; but the negative evidence, that is to say, the certainty of the non-inheritance of mutilations in a vast number of cases (see Darwin: "Variation of Plants and Animals under Domestication," ii. 23), is so overpowering, that it may still be reasonable to look upon the former as no more than a collection of coincidences. The earliest instance that I know of, that seems worthy of serious consideration, is that of Dr. Brown-Séquard's epileptic guinea- pigs, because it admits of verification; but this, if I understand his account rightly (Proceedings of Royal Society, x. 297), is open to some objection. It appears that Dr. Brown-Séquard found, during his researches into the cause of epilepsy, that, by a particular operation on the spinal cords of guinea-pigs, he could induce a convulsive disease very much like epilepsy. He operated upon many guinea-pigs, and kept them apparently apart from the rest of his stock, and noticed that their young were at times attacked with "epileptiform" convulsions, while the young of the rest of his stock never were; hence he concludes that the artificially induced epilepsy was transmitted hereditarily. My objection to this conclusion is, that if persons were brought up from childhood in a ward of epileptic patients, they would certainly acquire a tendency to epileptiform seizures by the mere effect of imitation. It is notorious that many an epileptic person has had his fits first brought on by witnessing the epileptic seizure of another. This, however, may be an unfounded objection, due, as has just been remarked, to misapprehension of an experiment, whose details deserve a fuller description. It is much to be regretted, that two subsequent memoirs, read by Dr. Brown-Séquard at the British Association in 1870, do not appear to have been published; their titles only are to be found in its Journal (p. 134). But he has communicated a most important *résumé* of other results to the *Lancet* (Jan. 1875, p. 7) regarding the inheritance of certain purely physical effects that were produced on the parent guinea-pigs by nerve-mutilation, and their occurrence in the offspring, in the same order in which they had appeared in the parents.

A special cause may be assigned for the effects of disuse in causing hereditary atrophy of the disused parts. It has already been shown that all exceptionally developed organs tend to deteriorate; consequently those that are not protected by selection will dwindle. The level of muscular efficiency in the wing of a strongly flying bird is like the level of water in the leaky vessel of a Danaid,[7] only secured to the race by constant effort, so to speak; let the effort be relaxed ever so little, and the level immediately falls.

In addition to much else that might be said in disparagement of evidence on which overmuch reliance has hitherto been put, we should recollect that it is hazardous to adduce the very gradual adaptation of a race to changed conditions as a proof that acquired habits are hereditarily transmitted, because when several generations elapse before any appreciable result can be observed, selection will have had many opportunities of operating. It is indeed hard to find evidence of the power of the personal structure to react upon the sexual elements that is not open to serious objection. That which appears the most trustworthy, lies almost wholly in the direction of nerve changes, as shown by the inherited habits of tameness, pointing in dogs, and the like, and the results of Brown-Séquard.

The conclusion to be drawn from the foregoing arguments is, that we might almost reserve our belief that the structural cells can react on the sexual elements at all, and we may be confident that at the most they do so in a very faint degree; in other words, that acquired modifications are barely, if at all, *inherited*, in the correct sense of that word. If they were not heritable, then the second group of cases would vanish, and we should be absolved from all further trouble about them; but if they exist, in however faint a degree, a complete theory of heredity must account for them. I propose, as already stated, to accept the supposition of their being faintly heritable, and to account for them by a modification of Pangenesis. Each cell may be supposed to throw off a few germs that find their way into the circulation, and thereby to acquire a chance of occasionally finding their way to the sexual elements, and of becoming naturalised among them. In illustration of this process, we may recur to political metaphor, and imagine the stirp to be represented by some country, and the germs by its inhabitants. We know that, in every country, travellers

[7] The fifty daughters of Danaus. Forty-nine of them cut off the heads of their husbands and were eternally condemned to fill a leaky vessel with water in the lower world .

from other nations occasionally find a place, which they can fill more suitably than at their own homes or elsewhere, and they become settlers. The population of the country may be as highly organised as it is needful to consider the sexual elements to be; every trade and profession may seem to be full; and yet the stranger obtains a lodgment, either through superiority or luck. He may displace one of the native-born inhabitants, or he may find an unoccupied corner which he can fill; anyhow, as a matter of fact, he becomes a permanent citizen.

The hypothesis of organic units enables us to specify with much clearness the curiously circuitous relation which connects the offspring with its parents.[8] The idea of its being one of direct descent, in the common acceptation of that vague phrase, is wholly untenable, and is the chief cause why most persons seem perplexed at the appearance of capriciousness in hereditary transmission. The stirp of the child may be considered to have descended directly from a part of the stirps of each of its parents, but then the personal structure of the child is no more than an imperfect representation of his own stirp, and the personal structure of each of the parents is no more than an imperfect representation of each of their own stirps. The political analogy to the common, but false, idea of the filial relationship is that which connects colonists to their parent nations: the relationship, according to the views in this memoir, is much more circuitous and feeble; it resembles that which connects the *representative government* of the colony with the *representative governments* of the parent nations. This, at least, is a first approximation: the second approximation consists in making allowance for the small power that exists, of transmitting acquired peculiarities; that is, for the power of the personal structure to react upon the sexual elements, and thereby upon the future stirp. To effect this, the analogy may be revised by supposing the governments of the parent states to have the power of nominating a certain proportion of the colonists.

It now remains to summarise briefly. I began by showing that certain postulates were admitted by most biologists, and that they gave a firm base whereon to develop a theory of heredity. By these, and by what appear to be their necessary consequences, I explained the object of double parentage, and

[8] Galton: I endeavoured to explain this in a paper, Proc. Royal Soc., 1872, p. 394. [Galton 1872c, see chapter 9, p. 79.]

therefore of sex. Then I dwelt on the restless movements of the germs in the stirp and the variety of their attractions and repulsions. Next I explained how it arose, that brothers or sisters were often very dissimilar; also, on other grounds, why twins derived from the same primary stirp were either very much alike or extraordinarily contrasted (this being a fact that had resulted from inquiries of my own). Then, I spoke of individual variation. Then, I argued that the developed part of the stirp was almost sterile, and that it was from the undeveloped residue that the sexual elements were derived. By this, I explained the almost complete non-transmission of acquired modifications; also the occasional deficiency in the offspring, of qualities for which the parent had been exceptionally remarkable, and for certain diseases skipping alternate generations. The theory was proposed that the successive segmentations of the stirp were not perfectly clean and precise, but that each structure included many alien germs, whereby the progeny of all the contents of the residue of the stirp were distributed over the body. This accounted for much that Pangenesis overaccounted for, and was free from objections raised against the latter.

The assumed evidence that structural changes reacted on the sexual elements was then discussed, and it was pointed out that certain changes were really collateral which had been commonly thought to be effected by inheritance. Some of the evidence that structural changes might react on the sexual elements was, however, accepted, and to account for its existence, a modification of Pangenesis was adopted; each nascent cell being supposed to throw off germs which occasionally found their way into the circulation, and ultimately obtained a lodgment in the already constituted sexual elements; this process being therefore independent of and subordinate to the causes which were supposed mainly to govern heredity. Finally, the exact relationship was defined, which connects the parents with their offspring.

Chapter 17

Review of Ribot on Heredity

Heredity; a Psychological Study of its Phenomena, Laws, Causes, and Consequences. From the French of Th. Ribot, author of "Contemporary English Psychology." (London: Henry S. King & Co., 1875.)

It may be affirmed with much truth that if we wish to learn what pursuit ranks highest in public opinion, we shall find it in the career of those men to whom statues are erected by public subscription. It happened that the writer of these lines not long since revisited Cambridge, where, as he walked admiringly among the many new improvements, his eyes fell upon a recently erected bronze statue. It was the only out-of-door statue in the whole town; it occupied a commanding position in the market-place, hard by the University Church, and only a few steps from being in full sight of the Senate House. He walked reverently up to it, pondering as he went as to the manner of the man whose memory it so proudly perpetuated, and lo! it was Mr. Jonas Webb of Babraham, the famous breeder of Southdown sheep. The erection of this statue by the agriculturists of a county in whose capital a great university happens to be located, is worthy of note. It expresses their genuine appreciation of the practical application of the laws of heredity to all descriptions of farm produce, and it may be accepted as an omen that the time is near when the study of those laws and of their logical consequences shall permeate the philosophy of the university. It must do so, because there is no branch of science which refers to bodily structure or to mental aptitudes, neither is there any theological doctrine in which the theory of heredity, either directly or as one of the principal agents in evolution, can hereafter be left out of consideration.

In the course of formation of every science there has always been an embryonic or prescientific period. Nothing then existed but detached pieces of evidence, of an unsatisfactory kind, laxly discussed and explained by wild hypotheses. But, at length, the methods of science succeeded in catching with a firm grip some of the loose materials, then more was seized, and so, with an ever-increasing rapidity of conquest, the whole of them became gathered together within the pale of law. Heredity has at the present time developed into a science; much is definitely established, and many questions seem to require for their solution little more than direct experiment or the simple but careful collection of statistical facts. There is consequently some need of a work that shall concisely and clearly set forth what is already known, and what are the undecided questions which most urgently call for solution and might at the same time be solved by any person, who chose to devote a fair amount of intelligent and steady work to the purpose.

M. Ribot's book does not do this; it is not a work on a level with the present knowledge, but it takes us back to the prescientific stage of heredity. It again brings to the light old anecdotes of questionable value, and again treats with seriousness, hypotheses that have become obsolete. Speaking generally, the work is that of a partially informed and very speculative writer, and by no means that of a man of science. It is written in a somewhat pretentious style, which has the effect of making the reader believe that some great discovery is about to be announced, and of fixing his attention until he reaches the end, when the deferred hope proves never to be realised. As examples of the kind of information which he freely accepts as evidence—among the illustrations of longevity, we are told that "a collier in Scotland prolonged his hard and dreary existence over one hundred and thirty-three years." We next have, as an example of exceedingly acute sense, a story extracted from Prosper Lucas, who was much too credulous of wonderful stories, of "Hirsch Daenemarck, a Polish Jew, who about the year 1840 travelled over Europe, showing by decisive experiments that he could read in a closed book any page or line that might be desired;" and of his son, aged ten, who "possessed this same faculty in perhaps a more remarkable degree." Curiously enough, I happen to know something about this very case, which was mentioned to me two years ago as an avowed instance of extraordinary memory. The subject of hereditary memory was and is of interest to me, and I therefore wrote to a very eminent and learned Jew, to whom I was referred for information. His reply lies before me: I do not repeat the names in his letter, as I did not ask permission to do so. This is an extract from it: "The feat to which you allude was performed by a Jewish rabbi, whose name, I think, was Hirsch Norwegen, who was popularly called the ' Sihas-Pole ' —i. e., the Talmud Pole (' Sihas ' being composed of the initial letters of the Hebrew words meaning 'the six sections' of the Talmud), and who, travelling through the principal parts of Europe about the year 1848, astonished even such men as —— in Berlin, —— in Prague, and ——- in Padua.

He was not only able to tell the words which a pin thrust through one leaf in any part of the Talmud would pass on the next, but on any number of subsequent leaves." In fact, he had learnt the enormous work (thirty-six volumes) more or less by heart, through the aid of a local as well as verbal memory of wonderful power, devoted to that end only. My correspondent gave me particulars of another instance of extraordinary memory of the same kind that existed in his own family. His father, "when he was seven years old, could say by heart the whole of the Pentateuch in Hebrew, verse by verse, together with the remarks of the principal commentators, Farihi, Ebn Ezra, and Rash-barn; and throughout life—he died aged seventy-seven—his knowledge of the vast Talmudical and Rabbinical literature was such that he was constantly appealed to for pointing out the sources of obscure references or allusions; and, in fact, he never seemed to forget anything—whether places, persons, facts, or ideas—with which he had once become acquainted." I have reason to believe that a powerful memory, exact in all matters of detail, is a characteristic of the Jewish race. M. Ribot says there is a lack of evidence to prove the heredity of strong memory; on the contrary, I find it abundant. It existed, as we have just been informed, in the family of Hirsch Daenemarck, and it exists in the family of my correspondent. But to proceed with M. Ribot's book. He quotes Le Vailliant on the half-breed children of the Europeans and Hottentots, that the moral nature is always determined by the father. When the father is a Hottentot, "the child has always the good nature and gentle and kindly affection of the father," but, in the converse case, they have "the germs of all vices and unruly passions." (!) Again, he quotes, apparently with perfect approval, the opinion "that there is an invariable connexion between the heredity of physical resemblance and the heredity of moral resemblance." I can only say that I have been so struck by the number of cases in which the child who had the features of either parent had not the character, that I should hardly be surprised if they proved to be the more numerous; but I have never as yet gone statistically into this question. Then he indulges in some absurd views about likeness descending through opposite sexes, and quotes approvingly a belief that the son is more like to his mother, and, through her, to his grandfather, than he is to his father.

The inaccuracy and feebleness of his deductions is, in many instances, very striking. Here is one which is perfectly inexcusable in a writer on heredity; he is speaking of the transmission of acquired habits, and uses an often published anecdote to prove his case. He says:—

"Habit is defined to be an acquired disposition. We ask if any purely individual habits are transmitted ? Instances of this are cited. Girou de Buzareingnes observes that he had known a man who had the habit, when in bed, of lying on his hack and crossing the right leg over the left. One of his daughters had the same habit from birth."

The only meaning to be attached to this is, that the man had no special instinct to cross his legs, that from some cause or other he did so, that he acquired the habit of doing so, and that he transmitted this acquired habit by inheritance to his daughter. But what possible right has anyone to infer from the story, as it is told, that the man's habit was not just as instinctive as that of his daughter? Everybody who knows anything of heredity is well aware that one of the most interesting questions at the present time concerns the possibility of transmitting acquired habits. There are some few, very few, well-known instances of it in animals, but hardly any in man, while there are a vast number of other instances in which acquired habits are most assuredly not transmitted in any recognisable degree. The question is of extreme interest in its bearing upon the rate and direction of evolution, and therefore every bit of evidence about it deserves the closest scrutiny; but M. Ribot passes complacently on, careless and unconscious.

It is necessary to draw serious attention to the large amount of unacknowledged plagiarism which characterises this book. M. Ribot has been immensely indebted for its general design, and for very many facts, to the well-known work of

Dr. Prosper Lucas, *Hérédité Naturelle* as the reader will sufficiently recognise by comparing the two tables of contents, but I myself am aggrieved yet more directly. I find the tables and genealogies that I had compiled, after very considerable research and sifting, and which I published in *Hereditary Genius*, appropriated without a word of acknowledgment. They are clipped and condensed, and a trifling number of names are varied, but that is all, and M. Ribot thinks fit to give this plagiarised version of the families of the principal poets, painters, musicians, men of science and of literature, statesmen, and commanders, exactly as if they were the fruits of his own discrimination and research. Nay worse, he mentions in three separate cases out of the whole number of them, that the genealogy of those cases was given after me, thereby implying that I had nothing to do with the rest. It is the more vexatious because he shows himself incapable of making the most of the materials he has thus conveyed to his own use, as, for example, in his tables of maternal and paternal influence, where he quotes a few cases on either side merely as anecdotes, and does not attempt to work the subject quantitatively.

The book improves towards its close, because the topics with which that portion of it deals, are more in accordance with the bent of the author's mind. He develops with effect the views that have of late become familiar to English readers, of the large part played by unconscious cerebration in intellectual acts, and. in one of his best passages he ascribes genius (as I myself have lately done, in ignorance that M. Ribot had anticipated me) to a large development of that portion of the mind. He says:—

"The highest creations of the imagination spring from the unconscious. Every great inventor, artist, man of science, artificer, feels within him an inspiration, an involuntary invasion, as it were, coming out of the depths of his being, but which is, as has been said, impersonal. All that comes under consciousness is results and not processes. The difference between talent and genius is the difference between the conscious and unconscious. Artists, prophets, martyrs, mystics, all those who in any degree have felt the furor poeticus, have ever acknowledged their subjection to a higher power than their own egos and this power is the unconscious overlapping the submerged consciousness."

The word "talent" in the above is open to objection, because it is usually understood to mean an "instinctive gift," and instinctive motives are not necessarily "conscious." The phrase ought to run " between steady brain-work and genius." I may add, that a woman's intelligence appears to have a larger proportion of the unconscious element in it than the man's, for it is notorious that she frequently arrives at just conclusions, though the only reasons she is able to assign may be eminently illogical.

Much is said in the book about free will, but nothing worthy of note is advanced. There is also an eloquent passage about the decay of the Greek genius, which is ascribed to the effects of "nature," but unhappily, the author does not even profess to understand the meaning of that phrase. He says:—

"Clearly heredity has nothing to do with this decay; but then if it is transmitted to the next generation, and if, further, the same causes go on acting in the same direction, it is equally clear that heredity in turn becomes a cause of decay."

These "ifs" and the uncertain conclusion, and the general haze that overspreads the passage, are characteristic of the author's style of reasoning.

In conclusion, I would remark, that it is usually as profitless as it is an ungracious task to pick out the defects in a man's work. Both the critic who studies it for his own information, and the reader of his criticism want, or ought to want, nothing else than to learn all of sterling worth that it contains. But in the present instance, no choice seemed open to me but to find fault, for I laid down M. Ribot's volume after honestly reading every line of it, with a weary sense of many wasted hours.

Francis Galton.

Chapter 18

Gregarious and Slavish Instincts

I propose in this chapter[1] to discuss a curious and apparently anomalous group of base moral instincts and intellectual deficiencies, that are innate rather than acquired, by tracing their analogies in the world of brutes and examining the conditions through which they have been evolved. They are the slavish aptitudes from which the leaders of men are exempt, but which are characteristic elements in the disposition of ordinary persons. The vast majority of persons of our race have a natural tendency to shrink from the responsibility of standing and acting alone; they exalt the *vox populi* even when they know it to be the utterance of a mob of nobodies, into the *vox Dei*,[2] and they are willing slaves to tradition, authority, and custom. The intellectual deficiencies corresponding to these moral flaws are shown by the rareness of free and original thought as compared with the frequency and readiness with which men accept the opinions of those in authority as binding on their judgment. I shall endeavour to prove that the slavish aptitudes in man are a direct consequence of his gregarious nature, which itself is a result of the conditions both of his primeval barbarism and of the forms of his subsequent civilisation. My argument will be, that gregarious brute animals possess a want of self-reliance in a marked degree; that the conditions of the lives of these animals have made a want of self-reliance a necessity to them, and that by the law of natural selection the gregarious instincts and their accompanying slavish aptitudes have gradually become evolved. Then I shall

1 Galton 1883, 47-57. This was reprinted, with some changes, from Galton 1871b.

2 But see Galton 1907 for Galton's reconsideration of *vox populi* in the light of averaging.

argue that our remote ancestors have lived under parallel conditions, and that other causes peculiar to human society have acted up to the present day in the same direction, and that we have inherited the gregarious instincts and slavish aptitudes which have been needed under past circumstances, although in our advancing civilisation they are becoming of more harm than good to our race.

It was my fortune, in earlier life, to gain an intimate knowledge of certain classes of gregarious animals. The urgent need of the camel for the close companionship of his fellows was a never-exhausted topic of curious admiration to me during tedious days of travel across many North African deserts. I also happened to hear and read a great deal about the still more marked gregarious instincts of the llama; but the social animal into whose psychology I am conscious of having penetrated most thoroughly is the ox of the wild parts of western South Africa. It is necessary to insist upon the epithet "wild," because an ox of tamed parentage has different natural instincts; for instance, an English ox is far less gregarious than those I am about to describe, and affords a proportionately less valuable illustration to my argument. The oxen of which I speak belonged to the Damaras, and none of the ancestry of these cattle had ever been broken to harness. They were watched from a distance during the day, as they roamed about the open country, and at night they were driven with cries to enclosures, into which they rushed much like a body of terrified wild animals driven by huntsmen into a trap. Their scared temper was such as to make it impossible to lay hold of them by other means than by driving the whole herd into a clump, and lassoing the leg of the animal it was desired to seize, and throwing him to the ground with dexterous force. With oxen and cows of this description, whose nature is no doubt shared by the bulls, I spent more than a year in the closest companionship.

I had nearly a hundred of the beasts broken in for the waggon, for packs, and for the saddle. I travelled an entire journey of exploration on the back of one of them, with others by my side, either labouring at their tasks or walking at leisure; and with others again who were wholly unbroken, and who served the purpose of an itinerant larder. At night, when there had been no time to erect an enclosure to hold them, I lay down in their midst, and it was interesting to observe how readily they then availed themselves of the neighbourhood of the camp fire and of man, conscious of the protection they afforded from prowling carnivora, whose cries and roars, now distant, now near, continually broke upon the stillness. These opportunities of studying the disposition of such peculiar cattle were not wasted upon me. I had only too much leisure to think about them, and the habits of the animals strongly attracted my curiosity. The better I understood them, the more complex and worthy of study did their minds appear to be. But I am now concerned only with their blind gregarious instincts, which are conspicuously distinct from the ordinary social desires. In the latter they are deficient; thus they are not amiable to one another, but show on the whole more expressions of spite and disgust than of forbearance or fondness. They do not suffer from an ennui, which society can remove, because their coarse feeding and their ruminant habits make them somewhat stolid. Neither can they love society, as monkeys do, for the opportunities it affords of a fuller and more varied life, because they remain self-absorbed in the middle of their herd, while the monkeys revel together in frolics, scrambles, fights, loves, and chatterings. Yet although the ox has so little affection for, or individual interest in, his fellows, he cannot endure even a momentary severance from his herd. If he be separated from it by stratagem or force, he exhibits every sign of mental agony; he strives with all his might to get back again, and when he succeeds, he plunges into its middle to bathe his whole body with the comfort of closest companionship. This passionate terror at segregation is a convenience to the herdsman, who may rest assured in the darkness or in the mist that the whole herd is safe whenever he can get a glimpse of a single ox. It is also the cause of great inconvenience to the traveller in ox-waggons, who constantly feels himself in a position towards his oxen like that of a host to a company of bashful gentlemen at the time when he is trying to get them to move from the drawing-room to the dinner-table, and no one will go first, but every one backs and gives place to his neighbour. The traveller finds great difficulty in procuring animals capable of acting the part of fore-oxen to his team, the ordinary members of the wild herd being wholly unfitted by nature to move in so prominent and isolated a position, even though, as is the custom, a boy is always in front to persuade or pull them onwards. Therefore, a good fore-ox is an animal of an exceptionally independent disposition. Men who break in wild cattle for harness watch assiduously for those who show a self-reliant nature, by grazing apart or ahead of the rest, and these they break in for fore-oxen. The other cattle may be indifferently devoted to ordinary harness purposes, or to slaughter; but the born leaders are far too rare to be used for any less distinguished service than that which they alone are capable of fulfilling. But a still more exceptional degree of merit may sometimes be met with among the many thousands of Damara cattle. It is possible to find an ox who may be ridden, not indeed as freely as a horse, for I have never heard of a feat like that, but at all events wholly apart from the companionship of others; and an accomplished rider will even succeed in urging him out at a trot from the very middle of his fellows. With respect to the negative side of the scale, though I do not recollect definite instances, I can recall general impressions of oxen showing a deficiency from the average ox standard of self-reliance, about equal to the excess of that quality found in ordinary fore-oxen. Thus I recollect there were some cattle of a peculiarly centripetal instinct, who ran more madly than the rest into the middle of the herd when they were frightened; and I have no reason to doubt from general recollections that the law of deviation from an average would be as applicable to independence of character among cattle as one might expect it theoretically to be. The conclusion to which we are driven is, that few of the Damara cattle have enough originality and independence of disposition to pass unaided through their daily risks in a tolerably comfortable manner. They are essentially slavish, and seek no better lot than to be led by any one of their number who has enough self-reliance to accept that position. No ox ever dares to act contrary to the rest of the herd, but he accepts their common determination as an authority binding on his conscience.

An incapacity of relying on oneself and a faith in others are precisely the conditions that compel brutes to congregate and live in herds; and, again, it is essential to their safety in a country infested by large carnivora, that they should keep closely together in herds. No ox grazing alone could live for many days unless he were protected, far more assiduously and closely than is possible to barbarians. The Damara owners confide perhaps 200 cattle to a couple of half-starved youths, who pass their time in dozing or in grubbing up roots to eat. The owners know that it is hopeless to protect the herd from lions, so they leave it to take its chance; and as regards human marauders they equally know that the largest number of cattle watchers they could spare could make no adequate resistance to an attack; they therefore do not send more than two, who are enough to run home and give the alarm to the whole male population of the tribe to run in arms on the tracks of their plundered property. Consequently, as I began by saying, the cattle have to take care of themselves against the wild beasts, and they would infallibly be destroyed by them if they had not safeguards of their own, which are not easily to be appreciated at first sight at their full value. We shall understand them better by considering the precise nature of the danger that an ox runs. When he is alone it is not simply that he is too defenceless, but that he is easily surprised. A crouching lion fears cattle who turn boldly upon him, and he does so with reason. The horns of an ox or antelope are able to make an ugly wound in the paw or chest of a springing beast when he receives its thrust in the same way that an over-eager pugilist meets his adversary's "counter" hit. Hence it is that a cow who has calved by the wayside, and has been temporarily abandoned by the caravan, is never seized by lions. The incident frequently occurs, and as frequently are the cow and calf eventually brought safe to the camp; and yet there is usually evidence in footprints of her having sustained a regular siege from the wild beasts; but she is so restless and eager for the safety of her young that no beast of prey can approach her unawares. This state of exaltation is of course exceptional; cattle are obliged in their ordinary course of life to spend a considerable part of the day with their heads buried in the grass, where they can neither see nor smell what is about them. A still larger part of their time must be spent in placid rumination, during which they cannot possibly be on the alert. But a herd of such animals, when considered as a whole, is always on the alert; at almost every moment some eyes, ears, and noses will command all approaches, and the start or cry of alarm of a single beast is a signal to all his companions. To live gregariously is to become a fibre in a vast sentient web overspreading many acres; it is to become the possessor of faculties always awake, of eyes that see in all directions, of ears and nostrils that explore a broad belt of air; it is also to become the occupier of every bit of vantage ground whence the approach of a wild beast might be overlooked. The protective senses of each individual who chooses to live in companionship are multiplied by a large factor, and he thereby receives a maximum of security at a minimum cost of restlessness. When we isolate an animal who has been accustomed to a gregarious life, we take away his sense of protection, for he feels himself exposed to danger from every part of the circle around him, except the one point on which his attention is momentarily fixed; and he knows that disaster may easily creep up to him from behind. Consequently his glance is restless and anxious, and is turned in succession to different quarters; his movements are hurried and agitated, and he becomes a prey to the extremest terror. There can be no room for doubt that it is suitable to the well-being of cattle in a country infested with beasts of prey to live in close companionship, and being suitable, it follows from the law of natural selection that the development of gregarious and therefore of slavish instincts must be favoured in such cattle. It also follows from the same law that the degree in which those instincts are developed is on the whole the most conducive to their safety. If they were more gregarious they would crowd so closely as to interfere with each other when grazing the scattered pasture of Damara land; if less gregarious, they would be too widely scattered to keep a sufficient watch against the wild beasts.

I now proceed to consider more particularly why the range of deviation from the average is such that we find about one ox out of fifty to possess sufficient independence of character to serve as a pretty good fore-ox. Why is it not one in five or one in five hundred?[3] The reason undoubtedly is that natural selection tends to give but one leader to each suitably-sized herd, and to repress superabundant leaders. There is a certain size of herd most suitable to the geographical and other conditions of the country; it must not be too large, or the scattered puddles which form their only watering-places for a great part of the year would not suffice; and there are similar drawbacks in respect to pasture. It must not be too small, or it would

[3] Galton's original formulation, in Galton 1871b, is more suggestive: "I will now put a question on the why and wherefore of all this, of the same form as might be put in respect to any strictly measurable character, such as human stature: for instance, we might say, why has such and such a race an average height of 5 feet 7 inches and why, again, is the deviation from that average of such a magnitude that one-twentieth part of the population exceeds 5 feet 10 inches. The inquiry I will now put in respect to the gregariousness of cattle falls into precisely the same shape. How is it that their self-reliance is, on the average, so low; and, again, how is it that the deviation from that average is such as to allow of the appearance of about one animal in fifty having the capabilities of a good fore-ox?"

be comparatively insecure; thus a troop of five animals is far more easy to be approached by a stalking huntsman than one of twenty, and the latter than one of a hundred. We have seen that it is the oxen who graze apart, as well as those who lead the herd, who are recognised by the trainers of cattle as gifted with enough independence of character to become fore-oxen. They are even preferred to the actual leaders of the herd; they dare to move more alone, and therefore their independence is undoubted. The leaders are safe enough from lions, because their flanks and rear are guarded by their followers; but each of those who graze apart, and who represent the superabundant supply of self-reliant animals, have one flank and the rear exposed, and it is precisely these whom the lions take. Looking at the matter in a broad way, we may justly assert that wild beasts trim and prune every herd into compactness, and tend to reduce it into a closely-united body with a single well-protected leader. That the development of independence of character in cattle is thus suppressed below its otherwise natural standard by the influence of wild beasts, is shown by the greater display of self-reliance among cattle whose ancestry for some generations have not been exposed to such danger.

What has been said about cattle, in relation to wild beasts, applies with more or less obvious modifications to barbarians in relation to their neighbours, but I insist on a close resemblance in the particular circumstance, that many savages are so unamiable and morose as to have hardly any object in associating together, besides that of mutual support. If we look at the inhabitants of the very same country as the oxen I have described, we shall find them congregated into multitudes of tribes, all more or less at war with one another. We shall find that few of these tribes are very small, and few very large, and that it is precisely those that are exceptionally large or small whose condition is the least stable. A very small tribe is sure to be overthrown, slaughtered, or driven into slavery by its more powerful neighbour. A very large tribe falls to pieces through its own unwieldiness, because, by the nature of things, it must be either deficient in centralisation or straitened in food, or both. A barbarian population is obliged to live dispersedly, since a square mile of land will support only a few hunters or shepherds; on the other hand, a barbarian government cannot be long maintained unless the chief is brought into frequent contact with his dependants, and this is geographically impossible when his tribe is so scattered as to cover a great extent of territory. The law of selection must discourage every race of barbarians which supplies self-reliant individuals in such large numbers as to cause tribes of moderate size to lose their blind desire of aggregation. It must equally discourage a breed that is incompetent to supply such men in sufficiently abundant ratio to the rest of the population to ensure the existence of tribes of not too large a size. It must not be supposed that gregarious instincts are equally important to all forms of savage life; but I hold, from what we know of the clannish fighting habits of our forefathers, that they were every whit as applicable to the earlier ancestors of our European stock as they are still to a large part of the black population of Africa.

There is, moreover, an extraordinary power of tyranny invested in the chiefs of tribes and nations of men, that so vastly outweighs the analogous power possessed by the leaders of animal herds as to rank as a special attribute of human society, eminently conducive to slavishness. If any brute in a herd makes itself obnoxious to the leader, the leader attacks him, and there is a free fight between the two, the other animals looking on the while. But if a man makes himself obnoxious to his chief, he is attacked, not by the chief single-handed, but by the overpowering force of his executive. The rebellious individual has to brave a disciplined host; there are spies who will report his doings, a local authority who will send a detachment of soldiers to drag him to trial; there are prisons ready built to hold him, civil authorities wielding legal powers of stripping him of all his possessions, and official executioners prepared to torture or kill him. The tyrannies under which men have lived, whether under rude barbarian chiefs, under the great despotisms of half-civilised Oriental countries, or under some of the more polished but little less severe governments of modern days must have had a frightful influence in eliminating independence of character from the human race. Think of Austria, of Naples and even of France under Napoleon III. It was stated[4] in 1870 that, according to papers found at the Tuileries, 26,642 persons had been arrested in France for political offences since 2nd December, 1851 and that 14,118 had been transported, exiled, or detained in prison.

I have already spoken in *Hereditary Genius* of the large effects of religious persecution in comparatively recent years, on the natural character of races, and shall not say more about

4 Galton: *Daily News*, 17th October, 1870.

it here; but it must not be omitted from the list of steady influences continuing through ancient historical times down, in some degree, to the present day, in destroying the self-reliant, and therefore the nobler races of men. I hold that the blind instincts evolved under these long-continued conditions have been ingrained into our breed, and that they are a bar to our enjoying the freedom which the forms of modern civilisation are otherwise capable of giving us. A really intelligent nation might be held together by far stronger forces than are derived from the purely gregarious instincts. A nation need not be a mob of slaves, clinging to one another through fear, and for the most part incapable of self-government, and begging to be led; but it might consist of vigorous self-reliant men, knit to one another by innumerable ties, into a strong, tense, and elastic organisation. The character of the corporate action of a nation in which each man judges for himself, might be expected to possess statistical constancy. It would be the expression of the dominant character of a large number of separate members of the same race, and ought therefore to be remarkably uniform. Fickleness of national character is principally due to the several members of the nation exercising no independent judgment, but allowing themselves to be led hither and thither by the successive journalists, orators, and sentimentalists who happen for the time to have the chance of directing them.

Our present natural dispositions make it impossible for us to attain the ideal standard of a nation of men all judging soberly for themselves, and therefore the slavishness of the mass of our countrymen, in morals and intellect, must be an admitted fact in all schemes of regenerative policy.

The hereditary taint due to the primeval barbarism of our race, and maintained by later influences, will have to be bred out of it before our descendants can rise to the position of free members of an intelligent society: and I may add that the most likely nest at the present time for self-reliant natures is to be found in States founded and maintained by emigrants.

Servility has its romantic side, in the utter devotion of a slave to the lightest wishes and the smallest comforts of his master, and in that of a loyal subject to those of his sovereign; but such devotion cannot be called a reasonable self-sacrifice; it is rather an abnegation of the trust imposed on man to use his best judgment, and to act in the way he thinks the wisest. Trust in authority is a trait of the character of children, of weakly women, and of the sick and infirm, but it is out of place among members of a thriving resolute community during the fifty or more years of their middle life. Those who have been born in a free country feel the atmosphere of a paternal government very oppressive. The hearty and earnest political and individual life which is found when every man has a continual sense of public responsibility, and knows that success depends on his own right judgment and exertion, is replaced under a despotism by an indolent reliance upon what its master may direct, and by a demoralising conviction that personal advancement is best secured by solicitations and favour.

Chapter 19

Rede Lecture, Cambridge, May 28, 1884

19.1 Cambridge Chronicle

There was a large attendance in the Senate House on Wednesday to hear the lecture by Mr. Francis Galton, M.A., F.R.S., upon "The Measurement of Human Faculty".[1] The gallery was well filled by undergraduates,[2] who displayed lively interest in the subject and took the lecturer's points with commendable promptness and energy.

The Vice Chancellor having introduced the lecturer, Mr. Galton said in his lecture he should discuss the possibility of taking more complete measurements of mind than had hitherto been practised. The powers of man were finite, and if finite, they were measurable. It was somewhat anomalous that the art of measuring the intellectual faculties (by examination) should have become so highly developed, while that of dealing with other faculties should have been so little practised, or even considered. To essay the measurement of man in his entirety was to be justified by exactly the same arguments as those by which special examinations—such as those of classics or mathematics—were justified, viz, that each measurement tests the adequacy of the previous education. But it was impossible to be sure of a man's total efficiency unless he had been measured in character, and physically, as well

[1] Report from the *Cambridge Chronicle and Journal*, Friday 30 May 1884, p. 4.

[2] The *Evening Mail* 1884/05/30, p. 3. described the audience as 'tolerably large' but 'not so large as is customary at this annual lecture' due to examinations.

as in intellect. The art of measuring the various human faculties now occupied the attention of many inquirers in this and other countries. In its entirety it was ill-suited to form the subject of a lecture. He should, therefore, confine himself to two of its branches, one of which was but little explored, and dealt with temperament and character, and the other was thoroughly methodised, and dealt with the simple characteristics of height, weight, strength, and the like. He might have a difficulty in convincing them that his views of the measurement of character were not visionary, but he was confident of being able to show the importance of similar measurements. First then, as regards character. Could they discover landmarks to serve as bases for a survey, or was it altogether too indefinite to admit of measurement? Was man, with his power of choice and freedom of will as different from a conscious machine that any proposal to measure his mind was based upon a fallacy? He had made a series of careful inquiries, the general result of which had led him to believe that the motives of the will were mainly normal, and that the character which shaped their conduct was a definite something, and, therefore, that it was reasonable to attempt to measure it. In making this measurement they must go upon the fundamental principles of intellectual examinations. Definite acts in respect to definite characters had alone to be measured. For this purpose he advocated the collection of data for comparative observations. To do this they had to fix upon some particular class of persons of similar age, sex, and social condition, and to find out what common incidents in their lives were most apt to make them betray their character. The selection made, it must be well experimented upon, for they must remember that observations of character were not yet developed and could only be accomplished after many trials. One of the most notable differences between man and man lay in their tempers. But these would display themselves by carefully watching the subject, and he felt sure if two or three experimenters were to act judiciously together as secret accomplices they would soon collect abundant statistics of conduct, and learn to probe character more clearly. By the use of these statistics and assuring themselves of the reality of emotions, they could form their calculations for generalizations. The first, that character ought to be measured by acts, and then that the practice of deliberately and methodically testing the characters of others, and of themselves, was not wholly fanciful but deserved experiment. Turning to the second portion of his subject, that dealing with physical measurement, he enlarged upon the importance of such tests as trying the senses of seeing and hearing, and noticing the bodily development. An exhibit would be in working order at the International Health Exhibition in the course of a few days, to be called the Anthropological Laboratory [sic], where all these tests could be applied. It was to be worked on a self-supporting basis, and it really seemed possible by judicious management to supply persons with the results of these tests, and keep a copy of them, for the very small sum of 3d. The value of simple measurements like these was very great, because it showed if their development was properly going on, and if defects were found in early life the opportunity might be afforded of remedying the hidden faults. The importance of the subject had been recognised in America, and the Johns Hopkins University had established a department for its consideration, where the defects of physical culture were shown and particular branches of gymnastics recommended as required. It had succeeded very well indeed. The value of simple measurements being indisputable, the cost of making them, as he had shown, being so trivial, and the value of the registers in any permanent institution being so obvious, it seemed strange that they should be neglected by any school or University. He was audacious enough to suggest the establishment of at least a plain anthropometric laboratory here at Cambridge (Applause.) He should consider it one of the best works of his life if the remarks he had had the privilege of addressing to that distinguished audience should lead to the introduction of a new acquirement in the University of the systematic but simple measurement of every undergraduate on his matriculation, and again before his taking a degree. (Laughter and applause.) The schools would gradually follow suit, and the growth and development of every member of the educated classes would be, for the future, systematically watched. Moreover, each person who had been at school and at at University would know where to collect data for his own life history whenever he desired to bring them together. It would be a step towards keeping a continuous record of the growth of life and disease, and must give a number of accurate records from which much valuable information might hereafter be deciphered. (Loud and prolonged applause.)

The Vice Chancellor expressed the thanks of those present to the lecturer, and the proceedings closed, Mr Galton being loudly cheered by the undergraduates as he left the building.

19.2 Manuscript Fragment

... I now turn[3] to the very opposite extreme of Anthropometry. that is to what in its many features is its simplest, best known, and longest established past. I cannot explain what I mean better than by reading the programme of an exhibit at the International Health Exhibition which will be in working order in a few days. I suggested it to the Council, who willingly adopted the idea, and I gave the instruments, and have arranged the methods of using them. It is to be called the Anthropometric Laboratory, where where a person may be measured in the following particulars. Eyesight, color sense, hearing, breathing capacity, strength, swiftness of blow, height, span of arms, weight. It is to be worked in a self-supporting basis, and it really seems possible by judicious management to supply any person with all these facts, for the very small sum of three pence.

The value of simple measurements like these, is that they show whether the physical development of the child or youth is proceeding normally. They draw attention to the existence of hidden faults in his rearing which would detract from his future efficiency as a man, and which if detected early, have a better chance of being palliated or overcome. Have there not been hundreds of thousands of cases of eyesight needlessly injured beyond repair, of lop-sided growth and stunted chest capacity, which if manifested by measurement would have been surely attended to in time, and remedied?

The necessity of periodical measurement is thoroughly recognised by those who have studied the science of Health, but it has not yet obtained that hold on popular opinion which it deserves, and which at some future time it will undoubtedly attain.

America, in some of her colleges has instituted a system of physical measurement, and is turning them to good account, going further in this direction than the public opinion in this country is probably prepared for. Thus in Johns Hopkins university, a new department has been established, called that of Physical Education and Hygiene, and a notice was issued last January, which I quote from its university circular, that

[3] From the surviving MS fragment GALTON/2/4/18/5.

"matriculate students will hereafter be required, in order to secure the degree of Bachelor of Arts, to take the course in Physical Culture". I am indebted for much information on this subject to Dr. Hartwell, who presides over that department. The students are not compulsorily measured but they present themselves of their own free will. They are bared of clothes and are so thoroughly examined that the schedule on which the results are entered, has some 50 separate headings, and corresponding blank spaces, all of which are filled up. Irregularities of muscular development are especially noted, and the student is advised as to some special form of gymnastic exercise. At intervals, the measurements are repeated and the gain or loss in each particular is recorded. (I should add, that numbers and not names, are entered in the chief records for the sake of privacy.) So far as I can judge from the numerous papers that have reached me, this elaborate system is very effective, is already popular and is growing in popularity. American experience shows how kindly students may take to a system even of minute measurement.

The rate of the annual increase of stature during youth, together with the records of weight, are easy tests of normal development. I will refer more especially to weight, because I happen recently to have had the opportunity of examining a series of weights of the same persons at different ages extending from early manhood to nearly the close of life. We are as yet so deficient in records of the weighings of the same individual during many successive years, that I believe I am almost the first person who has enjoyed such an opportunity. The particular series I speak of had an additional interest, from its including a large number of the best known men in English society during the last hundred years. The history of the collection is a curious and instructive one, and as follows.

In a well-known and fashionably frequented shop in St James' Street, there hang magnificent and ancient beam scales, one for the purpose of the shop and another for the use of the customers. Also, on a shelf at their side stand a row of ledgers, well indexed, ready to receive entries. The opportunity of being weighed was usually irresistible to those who entered the shop. Statesmen, generals, diplomatists, men of fashion daily sat in the scales, and the results were duly booked, and the records still are there. I mention these details to show how little averse people are from being weighed and measured. On the contrary, they like it. Nothing interests them more than particulars about themselves. The plainest people look at their own faces in the glass without displeasure.

The examination I made into the age-weights was confined to those of peers & some baronets, whose ages at the time of the weighings were easily found in the peerage. The results gave an excellent measure of the mischievous influence of the form of social life towards the end of the last century, when the phrase "as drunk as a lord" was in use, and too true as to fact. The oscillations in weight from year to year were more abrupt and large, then than now, and the times of life when men began to grow heavy and when they began to shrink were earlier than at present.

Anthropometric records are treated statistically to discover the efficiency of the nation as a whole, and in its various parts, and the direction in which it is changing, whether for the better or worse. We are still unable to speak with moderate precision about these things. The Anthropometric Committee of the British Association found great difficulty in collecting data, and their collection is sadly imperfect. There is great need for regular and systematic registration of physical measurements in a form available for statistical inquiry.

To show the use of preserving even the minor data, I will dwell for a moment on that of the colour of the hair and eyes, which might be thought at first sight to have no obvious bearing on the general efficiency of the nation. This is far from being the case. The British nation is partly a blend, partly a mosaic of very different types. The short black-haired, ancient British race unites imperfectly with the tall, fair-haired Danish or Scandinavian. Their union resembles what druggists call an emulsion, that is a mixture of oil and water so well shaken up that they form at the time an apparently homogeneous compound; but the combination is not durable. Leave the emulsion alone and after a while it will separate into its component elements. Types are stable, but the forms of their mongrel offspring are not, and whenever the external features of the old types reappear in something of their original purity, we may reasonably suppose that their inward characteristics are likely to be present also.

Whether it be as a race peculiarity or not, the color of the hair is related at least in America to certain immunities from, & susceptibilities to disease. This is shown by the medical and other statistics published by the American War Office in 1875, under the direction of Dr. Baxter. At the time of the war of the rebellion there, all male citizens of the United States between the ages of 20 and 45 were medically inspected with great minuteness to ascertain how many were fit for military service. About one quarter were unfit, and the diseases that incapacitated them are recorded in Dr. Baxter's book, in connection with other particulars, including the color of the hair. An analysis of between three hundred and thirty and three hundred & forty thousand cases of invalidism showed that the proportion of the light haired men in America who were unfit for service was larger than that of the dark haired men. The light haired men in America suffered the most from every form of disease except chronic rheumatism, which alone fixed itself in preference upon the dark-haired.

It follows that even the color of the hair is a proper subject for anthropometric record, much more may we feel assured that other obviously important personal data deserve measurement and registration.

The value of the simpler measurements being indisputable, the cost of making them so trifling, and the facility of registering them in any permanent institution being so obvious, it seems strange that they should be neglected at any school or university. I am audacious enough to suggest the establishment of at least a plain Anthropometric Laboratory here at Cambridge. I should consider it one of the best works of my life if the remarks I have the privilege of addressing publicly to this distinguished audience, were to lead to the introduction of a new requirement in the University, that of a systematic but simple measurement of every undergraduate, in his matriculation and again before taking his degree. I believe that the introduction of so harmless & interesting a novelty would be in reality a very important measure, not only in respect to its immediate object but because the schools would gradually follow suit, and the growth and development of every member of the educated classes would be in the future systematically observed. Moreover each person who had been at school and at the University would know where to collect data for his own life-history, whenever he desired to bring them together. It is not only personal feeling that I am about to express, but one that has been uttered quite as emphatically by not a few of the most eminent members of the medical profession, that one of the great desiderata of the time is a habit of keeping continuous records of growth, health and disease. The step I advocate moves entirely in that direction. Each life may be viewed as if it were a separate experiment, conducted on a system of trial-and-error, and costly in time, in effort, and in pain. The cost of the experiment is wasted if no accurate records are kept, whence the lessons it is able to teach may hereafter be deciphered.

FLY-LEAF

TO

THE RECORD OF FAMILY FACULTIES.

BY

FRANCIS GALTON, F.R.S.

MR. FRANCIS GALTON offers £500 in prizes to those British subjects resident in the United Kingdom who shall furnish him before May 15, 1884, with the best Extracts from their own Family Records.

These Extracts will be treated as confidential documents, to be used for statistical purposes only, the insertion of names of persons and places being required solely as a guarantee of authenticity and to enable Mr. Galton to communicate with the writers in cases where further question may be necessary.

The value of the Extracts will be estimated by the degree in which they seem likely to facilitate the scientific investigations described in the preface to the *Record of Family Faculties*.[1]

More especially:

(*a*) By including *every* direct ancestor who stands within the limits of kinship there specified.

(*b*) By including brief notices of the brothers and sisters (if any) of each of those ancestors. (Importance will be attached both to the *completeness* with which each family of brothers and sisters is described, and also to the *number* of persons so described.)

(*c*) By the character of the evidence upon which the information is based.

(*d*) By the clearness and conciseness with which the statements and remarks are made.

The Extracts must be legibly entered either in the tabular forms contained in a duplicate copy of the *Record of Family Faculties*, (into

[1] *Record of Family Faculties*, (consisting of tabular forms and directions for entering data, with an explanatory preface,) by Francis Galton, F.R.S. Macmillan & Co., 1884.

Works on similar subjects by the same author, and issued by the same publishers:—

Inquiries into Human Faculty and its Development, 1883.

English Men of Science, their Nature and their Nurture, 1874.

Hereditary Genius, its Laws and Consequences, 1869.

To these may be added the following, as one in which the author has taken part:—

Life History Album, prepared by the direction of the Collective Investigation Committee of the British Medical Association. Edited by F. Galton, Chairman of the Life History Sub-Committee, 1884.

Figure 19.1: Fly-leaf for Record of Family Faculties 1884 (recto)

FLY-LEAF TO THE RECORD OF FAMILY FACULTIES.

which, if more space is wanted, additional pages may be stitched,) or they may be written in any other book with pages of the same size as those of the *Record*, provided that the information be arranged in the same tabular form and order. (It will be obvious that uniformity in the arrangement of documents is of primary importance to those who examine and collate a large number of them.)

Each competitor must furnish the name and address of a referee of good social standing (magistrate, clergyman, lawyer, medical practitioner, &c.), who is personally acquainted with his family, and of whom inquiry may be made, if desired, as to the general trustworthiness of the competitor.

The Extracts must be sent prepaid and by post, addressed to FRANCIS GALTON, 42 Rutland Gate, London, S.W. It will be convenient if the letters "R.F.F." (Record of Family Faculties) be written in the left-hand corner of the parcel, below the address.

The examination will be conducted by the donor of the prizes, aided by competent examiners.

The value of the individual prizes cannot be fixed beforehand. No prize will, however, exceed £50, nor be less than £5, and £500 will, on the whole, be awarded.

A list of the gainers of the prizes will be posted to each of them. It will be published in one or more of the daily newspapers, also in at least one clerical, and one medical journal.

*** The above conditions are in lieu of those provisionally sketched out by MR. GALTON in the *Fortnightly Review* of August, 1883, for the purpose of eliciting suggestions, and which were subsequently submitted in a more elaborate form to many members of the medical profession. Their present shape is fixed in accordance with the balance of opinions elicited by these preliminaries, which was in favour of throwing them open to general competition, and not to medical men only, as at first intended.

MR. GALTON would add, that extracts from the Records of any family would be very acceptable to him, even such as may be too incomplete in the opinion of their authors to have the chance of gaining a prize, or which for any other reason should have the words, "not for competition," written upon them.

He sincerely trusts that his efforts to draw the attention of the public to the utility of family records, to induce them to rescue facts that are steadily gliding into oblivion, and to collect materials for serious scientific study, will be taken in good part and in the spirit in which they are made.

December, 1883.

Figure 19.2: Fly-leaf for Record of Family Faculties 1884 (verso)

Chapter 20

Galton's Royal Society Medal Speech, 1886.

The President (Professor George Gabriel Stokes) proceeded to deliver the anniversary address. ...[1]

The other Royal medal has been awarded to our Fellow, Mr. Francis Galton, for his statistical inquiries into biological phenomena. Mr. Galton is well known as an explorer and geographer, and his mind is singularly fertile in the devising of ingenious instruments for various objects. Many years ago he brought before us some remarkable experiments instituted with a view to test a particular biological theory in which rabbits of a pure variety were so connected with others of a different variety that the same blood circulated through both individuals, and the point to determine was whether this blood-relationship, in the most literal sense of the term, had any effect on the offspring.[2] Contrary to what the theory in question led us to regard as the more probable, the result proved to be negative. It is, however, in accordance with the rules of the award of the Royal medals, more especially the later investigations of Mr. Galton, in relation to vital statistics, that have been taken as the ground of the award. He has shown that by taking the average of a number of individuals having some condition in common, individual peculiarities apart from that condition are eliminated in the mean, and results are obtained which may be regarded as typical of that condition. One way of doing this is by his method of compound photographs. Thus we may obtain typical features of criminals of a particular kind, of consumptive persons,

---

[1] *The Times* 1886/12/01, 8.

[2] Galton 1871a.

and so forth. By adhering to the method of averages he has even succeeded in obtaining a mathematical expression, very closely verified in observation, connecting the mean deviation of some condition (such, for example, as stature) in a series of individuals, from the general mean of the same condition, with the mean deviations of the same condition in the relatives of those same individuals of different kinds, such as fathers, brothers, &c. Nor is the statistical method applicable to bodily characteristics alone. Mr. Galton has even extended it with remarkable ingenuity and originality to mental phenomena also. ...

At the anniversary dinner [of the Royal Society] in the evening, Mr. Francis Galton, in returning thanks for the toast of "The Gold Medallists," said "It is expected on these occasions that the recipient of a Royal medal should become egotistical and give some account of the influences that moulded his scientific career. I am very glad that I can do so briefly. To me, the first determining influence was a youthful spirit of adventure that goaded me to explore South Africa. I started in 1850 with abundant moral support from the Geographical Society, and succeeded, among other things, in being the first to traverse Damaraland, that country which has recently fallen under the protection of Germany and where I saw enough of savage races to give me material to think about all the rest of my life. Shortly after returning home one of the gold medals of the Geographical Society was awarded to me. I was elected a member of their council and then their honorary secretary, in which capacity I energetically occupied myself in furthering new expeditions, in planning instrumental equipments, and devising arts of travel. It was a heyday time for explorers. Almost the whole of Central Africa and Australia was unknown; historical Asiatic geography was little known and largely misunderstood. Steamships and new methods of sledging gave hope of an unlimited range to Arctic discovery. There was in every direction what Americans would call a great "boom" of travel. For the part I had played in all this, I had the honour of being elected a Fellow of the Royal Society in 1860. The ethnological aspects of geography now began to attract me more than the physical ones. It was about this time that the fact dawned on scientific men that the key to the origin of society among civilized nations and to many of their unexplained customs was to be found in the habits of contemporary barbarians. I can assure you, as a specialist in heredity, that I am not speaking without reason when I say that qualities which I seem to have inherited through two of my grandparents gradually yielded precedence to those that I certainly inherited from the other two. Recollect, please, that this medal is awarded to me for "statistical inquiries into biological phenomena." I can account fully both for the statistics and the biology. You must, please, allow me the pleasure of dissecting myself. On my father's side, I know of many most striking, some truly comic, instances of statistical proclivity. I have in my possession many pounds weight of ruled memoranda books, severally allotted to almost every conceivable house-hold purpose, which belonged to an aged female relative who died years ago. I also reckon at least five other remarkable instances of a love of tabulation within two degrees of kinship from myself. Again, as regards biology, I am sure there is a similarity between the form of the bent of my mind and that of my mother's father Dr. Erasmus Darwin. The resemblance chiefly lies in a strong disposition to generalize upon every-day matters that commonly pass unnoticed. I have myself attempted some of the very inquiries to which he had drawn attention, in complete unconsciousness that he had done so. It was owing to this hereditary cast of mind that I was well prepared to assimilate the theories of Charles Darwin when they first appeared in his "Origin of Species." Few can have been more profoundly influenced than I was by his publications. They enlarged the horizon of my ideas. I drew from them the breath of a fuller scientific life, and I owe more of my later scientific impulses to the influences of Charles Darwin than I can easily express. I rarely approached his genial presence without an almost overwhelming sense of devotion and reverence, and I valued his encouragement and approbation more, perhaps, than that of the whole world besides. This is the simple outline of my scientific history. I conclude by assuring you of the pleasure I have in the thought that this medal is a new bond, literally a golden bond, to connect me even more firmly than before with the Royal Society.

Chapter 21

Egypt, Soudan and Syria, 1845-6

F. Galton written from memory 1885.[1]

After my Father's death, Oct^{br} 1844, finding I had a competent fortune and hating the idea of practising medicine, also being disheartened by the sense that the medical knowledge to which I had access was very lax and that its progress seemed barred—I don't do justice I know to the state of the case, but only describe my feelings fresh from the rigorous methods of proof at Cambridge—that I determined to give it up. My passion was for movement & travel & I ultimately started for the East in September (I think), 1845.[2] On going by diligence south from Paris, I found myself with Denham Cookes as a companion. He was charming & full of anecdote & fun; we travelled together to Avignon, where we stayed some days & there I left him. He was killed at a steeplechase in Florence soon after.[3]

Stopped at Malta where Temple Frere was with his Uncle Hookham Frere.[4] On reaching Alexandria, (or was it en route

[1] GALTON/1/3/1/1. Transcribed directly from the manuscript. See also Pearson 1914, 199–205.

[2] It was no earlier than October, according to his letter from Harry Hallam on 1845/10/03.

[3] In fact, two years later, in April 1847. For Cookes, Denham Charles Johnstone, see page 354.

[4] Robert Temple Frere (1820–1886) and John Hookham Frere (1769–1846). "Robert Frere, one of my fellow-pupils when with Professor Partridge, became through marriage in later years a managing partner in a very old and eminent firm of wine merchants. They had supplied George iv. with his brandy and the like. He told me that the books of the firm showed that every class of wine had in its turn

after Malta?) I met with Montagu Boulton & his travelling companion, Hedworth Barclay, intending to go up the Nile. They having just toured in Greece. We went together to Cairo in barges, towed up the Nile by a tug—a most luxurious & charming night. At Cairo we three agreed to make a party together up the Nile. Barclay had a courier, a Greek, Christo, who would act as cook. Boulton had a very smart courier, Evard by name, who had been once groom of the chamber, he said, to Lady Jersey. He would be butler & I undertook to engage a dragoman as my special servant, & he was Ali (Mohammed, Sureyah = the little).

We spent one day boar shooting, got no boar but wounded one & when among the tall reeds holding my gun vertically above my head, as the only chance of making way, a litter of wild pigs ran through at my very feet! I was quite helpless, could not get down my arms, but the sow did not pass me. We had to give a name to our boat & register it with its flag, & Boulton suggested an ibis with the motto "tutissimus,"[5] which we adopted. Barclay had a pointer for quail shooting.

My impressions of the Nile were those that so many have expressed. Especially the pleasure of living all day barefoot & only half dressed, & of waking oneself by a header into the river, clambering back by the rudder. We lived in style & state.[6] I think the awfulness of the Old Temples impressed me most at Carnac,[7] going among them alone by moonlight & the silence broken rarely by the jackal. The feeling was so strong that it nearly made me faint clean away.

A little after on the first cataract, when near Korosko, the stream being swift, we went as usual, by virtue of Barclay's firman, to impress men to tug our boat, but found they had all been already impressed by the owner of a small & dirty looking Egyptian boat, who they told us was a Bey. We went to

---

been favoured by the doctors."

5 'Most prudent'.

6 'In Southern Africa I never could walk barefooted; independently of the thorns, there was something in our state of health which made small wounds difficult to heal, and caused scratches in the foot and hand to fester. Our very Damaras could not travel even with their own sandals, much less could we leave off shoes entirely. I was the more surprised at this, as in previous travelling in North Africa I had become nearly independent of them. I recollect climbing Jebel Barkal, which is a well-known rugged hill, with very sharp stones in it, near the fourth cataract of the Nile, barefooted.' Galton 1853, 186. In Galton 1908b, Galton recalled badly cutting his feet on Jebel Barkal.

7 Luxor.

him & spoke impudently, like arrogant Britishers & discussed loudly in English together whether we should not pitch him into the river. He shortly astonished us by speaking perfect French & after a while discovered he was a much more interesting & important person than we had dreamt of. He was Arnaud,[8] a St Simonian exile, in service of Mehemet Ali, who had lately returned from Sennaar where he has been sent to look for gold. He invited us to his (mud) house, at which I was charmed. Perfectly simple, clean, matted, with a barometer & thermometer hung up and other scientific gear, books, &c., like a native philosopher. He then, after we had become friends, explained to us, that though he spoke English badly, he quite understood what we had said among ourselves when we first met him & made me feel very small indeed. However, we got on very well & made him talk of his travels & tell us of the country ahead. We had then no map & knew nothing hardly. He said why do you follow the English routine of just going to the 2nd cataract & returning? Cross the desert & go to Khartoum.

That sentence was a division of the ways in my subsequent life.[9] We caught at the idea, he discussed it & said that the chief of the Korosko desert was then actually at the place with camels, that he knew him & would send for him to us that afternoon or evening, when we might finally settle matters. We asked Arnaud to dinner, received him in the grand style, Evard doing his best, & gave our good friend and ourselves quite as much wine as was good for us. When in the midst of the carouse the door of the cabin opened, the cool air came in, & with the cool air, the dignified cold presence of the Sheikh, with the band of sand on his forehead, the mark of his having just prostrated himself in prayer. He did look disgusted, but we got over him & finally all was arranged. We were to start the very next afternoon.

We settled to leave the boat, her captain & crew under the charge of Bob, our Arab pipe boy, as our representative, who rose easily to the position & they had orders to take the boat to Wadi Halfa & await us there. This was in mid January, we expected to return, as we did, early in March. We got

---

8 Théodore Louis Joseph-Pons d'Arnaud (1811–1884).

9 Galton is referring to the fact that when he first approached the RGS with the idea of an expedition in 1849/50, his idea was to explore further down the White Nile, as Arnaud had done with limited success in 1839–42, in search of its sources. The RGS persuaded him to go to Lake Ngami instead.

off on camel back in the afternoon & encamped 3 miles from Korosko & next morning started fairly off. It was a desert, like the skeleton of the earth, with sand blown clean away from the bare stones, or lying here & there in drifts. Table topped hills. Evard had some sort of eruptive fever & was frightfully depressed & lamenting, then Boulton got it & bore bravely up. It was hard lines for them. The water 4 days from Korosko, the only wells on the route, was brack & undrinkable; that in our water skins was horrible with the taste of leather. The waste of desert was terrible, & the way was marked by bones of slaves & camels. Often a dead camel was desiccated; it looked fairly right, but when touched broke & crumbled into dust, all the inside was blown away, or eaten away by the ants leaving the skin & part of the bones. These desiccated bodies were so light, that I once held up what appeared to be half a camel when first seen & as it lay untouched.

Our guide, a son or nephew of the Great Sheikh, was a jovial gay fellow & we all became excellent friends. Others joined our caravan; a man, his wife, baby & donkey—just like Joseph's flight. Also another man on foot, with no possessions but an old French cuirassier's sword, wherewith he was going to join slave raids in Abyssinia.[10] In 8 days from Korosko, we reached Abu Hamed—the sight of the Nile most refreshing, but we soon tired of the midges & riverian air & were glad to travel on a little inland by camel to Berber. On the way we stopped at the 5th cataract, where we waded with our guns across the river among the many islands. At Berber the Pasha received us in state & gave us lemonade from his own limes & it seemed delightful. He also lodged us in a mud house & gave permission to hire a boat for Khartoum. The people were troublesome when we tried to start, & seized the rope & wanted to detain us. Barclay behaved with much pluck, cast off the rope & made the 2 or 3 men who were on board hoist the sail. We got away & after a little, the rest of the crew ran along the bank & swam to us, & we got off. It was quite a small 1-masted boat, cabin 4 feet high, cockroaches all about but we made shift well enough.

I recollect little of the sail to Khartoum, except the mud pyramids of Meroë by the way. At Khartoum we got (I suppose through the captain of our boat) a mud house facing the blue Nile across which the dust columns were seen in numbers dancing on the plain. We heard of the existence of a wonderful

[10] Galton took this sword with him to South-West Africa.

Frank, possibly an Inglese; so we went to see. We knocked & walked in & there was about the most magnificent physique of a man I have ever seen, half dressed in Arnaout costume,[11] looking quite wild, & it turned out to be Mansfield Parkyns recently arrived there after years in Abyssinia. He had been at Trinity Coll: as well as ourselves & having taken part in an awkward row, found it best to leave, & had travelled ever since. He put us in the way of all the "life" in Khartoum & introduced us to the greatest scoundrels I think, that could be found anywhere in a room. Men who were too rascally for the Levant or even Cairo. They were slavedealers, outlaws & I know not what else. Full of stories about how A had been poisoned by B, B having just left the room before the story was told &c. Parkyns with perfect sang-froid & with all his wits well about him, held his own unscathed in this blackguard Bohemia, not a bit sullied by it & much amused.[12]

We arranged with our Captain to take us all together a little way up the Nile. We could not spare time for more than a very few days. So up we went. The mournful character of the big, slow, marsh-like expanse of river was very depressing. The air was heavy & seemed to me pestiferous & I was heartily glad to get away from it. The hippopotami were in great numbers, I blazed at 40 different ones (at absurd distances though) in one day. Boulton went out one night with Parkyns & shot a poor cow by mistake for one. There was no perceptible current in the river. The offal & cook's messes that were thrown overboard each night when the boat anchored, hung about her all night & were still there in the morning, so that we had to send a man wading to a distance to fetch clean water. The river lapped over the sloping banks like a flood over a meadow. There were vast flights of flamingoes, &c. & the aspect of the river was weird and strangely melancholy.

We turned back short of the Shilluk country, & returning to Khartoum, where we dropped Parkyns sailed on to Metemneh. There we engaged camels to cross the Bayuda desert. It hardly ranks as a desert as there are many watering places, we

---

[11] Ethnic Albanian. This conflicts with the account of Parkyns' dress in Galton 1908b, unless one supposes he was wearing, say, an Albanian-style headdress with a leopard skin draped over his shoulder.

[12] 'In the beginning of June Mr. Parkyns was at Khartoum at the junction of the Blue and White Rivers, where he had been detained for some time by a severe illness; and in August he expected to proceed down the Nile on his way to Cairo.' *Enniskillen Chronicle and Erne Packet*, Thursday 27 August 1846

only took 2 or 3 days rations of water with us and travelled 14 & even 16 hours a day. I started equipped in native dress, just a white cloth wrapped round, with arm & shoulder bare. The effect was I got fearfully blistered by the sun, all my back & arm was covered with minute blisters side by side. It was fearfully painful at night for some days. We travelled late into the night & the tail of the Great Bear was the index of our 24 hour clock. We met Prince Pukler Muskaw (?spelling) by the way.[13] I saw nothing of any wells, for we camped at night away from them & the camel men fetched the water. The ground had not at all the utter desert look of the Korosko. Rain falls there periodically & there are plenty of shabby mimosa trees.

We were 6 days in getting to Merawi.[14] There we stopped a few days wondering at the white ants. Everything had to be laid on "angarebs", [see Figure 21.1]

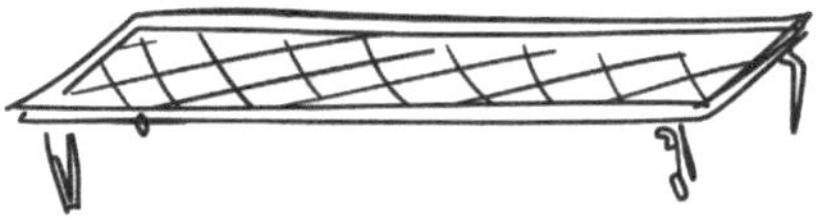

Figure 21.1: Angarebs.

frames with strips of hide across, & on legs, otherwise the white ants got at them. I went up Jebel Barkal[15] bare footed as a bravado, & the sharp edges of the schist like rocks severely punished my feet. There I got the vase, with what I now know was the representative of the God Bess upon it (given to the British Museum).[16] From Merawi we went a short 3 days ride across the desert to New Dongola where the Pasha was a much grander person than any hitherto seen. He had a review in our honor & mounted us on thoroughbred ponies with their queer Arab seats & with the cruel curbs. We all made a great mess of our riding with so unusual a seat, & if we touched the curb up went the plaguy ponies' heads, who were always

[13] Hermann, Fürst von Pückler-Muskau (1785–1871). Galton probably means that he saw Puckler-Musckau's graffiti at Jebel Barkal, since the German was in the region only in the late 1830s, and had already in 1844 published his travelogue, Pückler-Muskau 1845, covering that trip. In 1846 the Prince appears to have been in Europe.

[14] Merowe.

[15] The temple of Amun (Nubia), in Karima, Sudan.

[16] British Museum EA14957. Redware pottery vase decorated in the form of Bes. Excavated/Findspot: Dongola.

at a gallop or a sudden stop. The Pasha gave a monkey, to add to the two I had got at Berber & which were my constant companions in travel sitting on the camel with me or if not, with someone else.

From Dongola we rode along the left[17] bank of the Nile to Wadi Halfa, passing that wonderful Semneh. The Nile was then low & ran in a sluice between two low rocky banks that are under water at other times. It was so narrow that we thought we might throw a stone across it & tried hard to do so, but failed, some throws (not mine for I can't throw a bit) nearly succeeding. The river whenever we looked down upon it during our journey seemed totally unnavigable, seething among jutting rocks that were thickly set in its bed. At Wadi Halfa to our joy, we found our boat all right & Bob lording it with undisputed sway. He had actually ordered the Captain to be flogged for some offence, & the men obeyed Bob & flogged the Captain accordingly. Such a difference between the Berbers and the Egyptians. You can not strike a Berber but may flog as many Egyptians & beat them with sticks as much as you like. They are thoroughly slavish.

The voyage back, through March, became unpleasant from the Khamsin wind. It was uneventful except in the usual Nile experiences. At Cairo we hired a house in one of the quarters of the town with a big wooden key & lived there a week, conforming of course to the native ways of being in-doors by a specified & not late hour. Finally we separated. Barclay returned straight to England, Boulton by the short desert to Syria, & I not being particularly well, by steamer to Beyrout.

The awakening in the early morning when sailing along the shores of Syria & seeing the Holy Land for the first time, is one of the living pictures in my memory. Here my memory fails me. I was somehow in quarantine at Akka[18] & made great friends with the Pasha there. On the other hand I fancy I went straight to Beyrout. Such a change from Egypt. The people seemed so much less sedate & disagreeably go-ahead, and the verdure & hill slopes were so great a novelty to the eye. I lived rather stylishly, bought 2 good horses & a pony & jobbed a native groom, Ali remaining as my personal servant.

Furnished with introductions from my Akka friend, I stayed a night with the great Druse chief in his stronghold, who fêted me with distinction, believing evidently that I was a

[17] West.
[18] Acre.

much greater personage than I was, which rendered the stay embarrassing. I went to Damascus & boarded in the house of the English Doctor (Thompson). Thence as the heat was increasing I moved to Salahiyeh, where I took a house & set up an establishment in which figured my two Soudan monkeys & a pet ichneumon.[19] I lived a very Oriental life & became a fairly fluent talker in common Arabic, though nothing of a scholar. In fact, I am ashamed to say, I never read or even deciphered it fluently.

It was before I went to Salahiyeh & while still in Dr Thompson's house that faithful Ali was seized with dysentery. After an evening of parched skin & low delirium he died, with my money belt, that was under his charge, still round his body. We had the washers in & all the Moslem ceremonial duly attended to & I followed him to his grave, standing of course far off so as not to pollute. It was a great and serious loss. I was sincerely attached to him & condoned willingly heaps of small faults in regard to his great merits. One cold night in the desert when he & I were both chilled through, he pushed over me his rug. I did not know it till morning.

I got many rides from Salahiyeh & spent many pleasant afternoon hours in Arab caffées, sitting by the flowing waters. Colonel Churchill lived with much display near by, with his Syrian wife & I had a pleasant stay there. Finally, when the summer heats had passed I went to Lebanon & stayed a week with the Sheikh of Aden,[20] a right good fellow. The first morning I counted 97 flea bites on the right lower arm & up a little way above the elbow. There were Druse rows going on while I was there & we had to stand a brief attack, shots being fired & the house temporarily barricaded. Going thence to Tripoli I saw the most beautiful view on which my eyes have ever rested. It was of the Mediterranean through a gorge.

At Tripoli I did a foolish thing, viz. slept in the low marshy land & caught an ague that plagued me until I wholly lost it in 1851 in Africa. Riding along the shore towards Beyrout I met Boulton—a joyous meeting. We crossed & parted & I never saw him again. He went Eastwards & finally being an onlooker at the siege of Mooltan, with General Whish, took up a post of observation through a loophole in a deserted turret & when there a matchlock ball passed through his eye and brain. He was singularly gifted & amiable; ~~fond of~~ an epicurean in

[19] Egyptian Mongoose.
[20] Eden.

disposition, that is to say a philosophical pleasure seeker & of sterling merit.

At Beyrout I found my groom & horses had got into scrapes & I sold the latter. Being unwell with ague I felt unable then to ride to Jerusalem, so I took a place in a common collier sailing to Jaffa,[21] making myself supremely comfortable with rugs &c., on a cleaned corner of the deck. At Jaffa I found baggage camels, & in defiance of usage rode one into Jerusalem. The time when the Acre episode occurred & my stay in Mount Carmel has quite escaped me. It was there that Mr —— baby died & I performed some share in christening it just before its death. Also a Jesuit priest (as I believe) got hold of me & took great pains to convert me. Also I had a scramble at night to find, as I ultimately did, a wretched piece of humanity, a converted Jew, who had wandered about the hill & contrived to get himself into grief & lost himself & was become rather desperate when found.

At Jerusalem I planned an expedition, common enough now but then quite new, with one fatal exception of a year or two previous namely, to follow the valley of the Jordan all the way from Tiberias to the Dead Sea. Until Costigan's time (Mrs Bradshaw's brother of Leamington) from that of the Crusades, I believe there was no record of a Christian having attempted the journey owing to the wars of the tribes & the impossibility of getting safely from each to its neighbour. But a time of peace had set in and I availed myself of it. The plan was to get water skins at Jerusalem, take them on horseback to Tiberias, inflate & make a raft of them & on it to float down the Jordan. Starting from Jerusalem well escorted with spearmen & all mounted, including my native cook & I think one or two others, we ultimately slept at ——— overlooking the valley of the Jordan, half way along its course (there was a row at night & some of the horses tails were cut off in derision by the attackers). Thence I descended to the valley & rode up to Tiberias.

After a few days stay I started back, rigged out my raft just below the bridge where the Jordan issues from the lake, to the great amusement of the escort, who had orders to ride by the side, & off I floated. The stream was far too narrow & I got capsized twice. Then came a more serious misadventure for the current swirled in a narrow channel under overhanging boughs nearly touching the water & I was knocked off &

[21] Haifa.

got into difficulties. I soon saw that the raft project was not feasible at that season & took again to my horse. It was really a picturesque group. I had to ride in Arab head dress with a fillet & my men with their clump of long spears with ostrich feathers at the top looked very well indeed. After a while we came to a great Arab encampment, that of the Emir Ruabah whose sister, a relative of some kind of my own escorter, Sheikh Nair Abu Nasheer (of Jericho & thereabouts) had married. He was civil but wary & punctilious, & wherever I went I was watched. He had a quantity of old chain armour, beautiful Saracenic coats of mail. It was a somewhat uneasy visit to me & I was glad to be off.

We finally got to Jericho & thence to Jerusalem, I making various plans with my Sheikh for bringing a boat from Jaffa & with his permission navigating the Dead Sea, as Costigan had done, but poor fellow he had died in the act.[22] On returning to Jerusalem I found letters urging me to return home principally on account of some trust business for my sister Adèle. I went, hardly thinking it was a final parting with Syria, but so it was & the next year Lynch the American came, subsidised my Sheikh, surveyed the Jordan & Dead Sea very thoroughly & published the results in a big and valuable book, which by the way does not contain a word of allusion to myself, his predecessor.[23]

Francis Galton.

---

[22] Christopher Costigan, who died in 1835.

[23] Captain William Francis Lynch (1801—1865), who later served in the Confederate navy. He wrote *Narrative of the United States' Expedition to the River Jordan and the Dead Sea* (Philadelphia: Lea and Blanchard, 1849).

Figure 21.2: Galton's Sketch of Hedworth Barclay on the Nile, circa 1846. 'Barclay disagrees with McCulloch on the fixed duty! He remains unconvinced but determines on proceeding.' Restored.

Figure 21.3: Hedworth Barclay, Absorbed in Scott's Waverley. *By FG. Restored.*

Figure 21.4: Perhaps Arnaud Bey. Sketched by Galton on the Nile circa 1846. Restored.

Figure 21.5: Unknown subject, sketched by Galton circa 1846.

Figure 21.6: Impressions at Assouan, April 1846, by FG. Restored.

Figure 21.7: Throwing the Capo. Tomb Hieroglyphic, Sketched by FG in 1846.

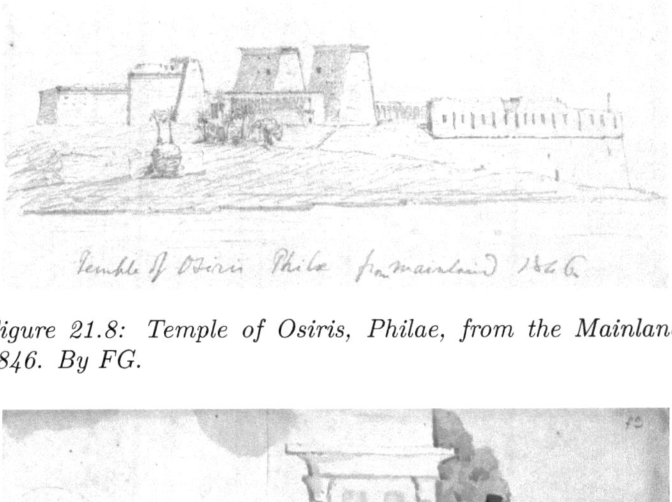

Figure 21.8: Temple of Osiris, Philae, from the Mainland, 1846. By FG.

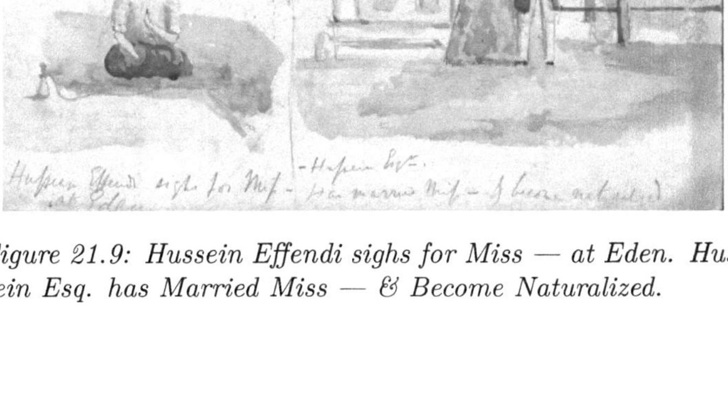

Figure 21.9: Hussein Effendi sighs for Miss — at Eden. Hussein Esq. has Married Miss — & Become Naturalized.

Chapter 22

Mental Fatigue, 1888[1]

The following paper was then read by the President:—

REMARKS ON REPLIES BY TEACHERS TO QUESTIONS RESPECTING MENTAL FATIGUE. BY FRANCIS GALTON, F.R.S., PRESIDENT.

The question of over fatigue of the brain in schools was discussed some months ago with much heat, and the arguments on either side were supported by experiences that so flatly contradicted each other, as to make it difficult to arrive at just conclusions. After the heat of discussion had somewhat cooled down, it happened that I was asked to occupy the chair at a meeting of the Educational Section of the Teachers' Guild, and while doing so I was much impressed by the eager and sustained attention of the large audience to the memoir read on that evening. It occurred to me that the Teachers' Guild might become a powerful instrument for the solution of statistical problems, if the interest of its intelligent members could be excited in inquiries bearing on Education, and if their good will and confidence could be gained. I determined to make a trial, and selected questions bearing on fatigue for the purpose. The Council of the Guild kindly assisted me by circulating my questions, together with a covering letter from their Vice Chairman, Dr. Morse.[2] The replies to those questions form the basis of the following remarks. Let me say at once, that I was somewhat disappointed in respect to the number and fullness of the replies—so much so, that I long hesitated to publish anything before supplementing them with other materials, to be gained gradually through my own observation, but having much else on hand, it seemed on the whole best to

[1] Galton 1888.

[2] Rev. Thomas Daniel Cox Morse (?–1895).

work off this matter at once, without admixture. I have 116 replies from teachers, many of large experience, concerning both themselves and their pupils, and as this is just sufficient to deserve a separate discussion, I shall not travel beyond the bounds of what may be called my brief. I will not enter upon other materials, and barely into the psycho-physics of fatigue, but shall merely set forward in an orderly way the statements contained in the 116 replies.

The objects of my questions were first to determine the signs and effects of incipient fatigue in as measureable a form as possible; for it is obviously most desirable to know what the tests of fatigue should be, in consequence of the contradictory opinions above alluded to. There ought to be no room for doubt as to whether the pupils in a particular school or class, and at a particular time, were or were not over fatigued. Secondly, I wished to hear from the teachers whether they had themselves ever broken down from over work, and what their own experiences might be concerning their pupils and friends. The actual questions are subjoined: numbers 1, 2, and 3 regard the person addressed; 4, 5, 6, regard their pupils and acquaintances.

1. What particular mental work can you perform easily, when your mind is fresh, that you find difficult or impossible when your mind is somewhat fatigued?

2. Has illness, due solely to mental overwork, independent of domestic anxiety and worry ever incapacitated you for more than a month at a time from ordinary school work? If so, give dates and symptoms. Do you consider your present health to be in any way affected by that illness?

3. Has experience discovered to you any warning signs, bodily or mental, distinct or obscure, of the imminent approach of mental fatigue, other than the growing sense of becoming fatigued? If so describe them.

4. What particular intellectual work do you find your pupils perform with ease when their minds are fresh, in which they fail more or less when they are mentally fatigued, even though they are still interested in their work?

5. Have you known cases of more or less serious prostration from mental overwork, as distinguished from the effects of domestic or other anxiety? If so, give initials and dates, and a very brief notice of the severity and duration of the illness?

6. Has experience discovered to you any warning signs of imminent mental fatigue among over zealous pupils?

The upshot of the replies to the questions is as follows:—

General Aspect.—Experienced teachers place most dependence on the general aspect of their classes, due to a variety of small indications, such as jaded expression and abnormal skin colour. They more especially speak of a strange look in the eye, which is variously described as dazed, weary, fixed, or lack lustre, as being a peculiarly characteristic indication that work should be slackened at once.

Nervous Irregularities.—Restlessness appears to be the commonest sign of partial fatigue: that is, of the attention being wearied while the muscles are craving to be employed. I may here for one moment break my plan of not travelling beyond my brief by alluding to a short account I wrote in "Nature" three years ago, Vol. xxxii, p. 174, but signed only with my initials, entitled "Measure of Fidget," describing how I had succeeded in counting the varying rate of fidget of a section of a large audience during the reading of a wearisome memoir. I have since frequently tried this method; it is an amusing way of passing an otherwise dull evening, but in drawing conclusions from the number of movements the average age of the audience and their habits of thought have to be taken into account. Children are extraordinarily mobile, and those adults who are little accustomed to concentrate their attention, are rarely still except when spellbound by eloquence. On the other hand I have frequently noticed at meetings of the Royal Society, that as many of the persons present as I could hold in a glance, were all as rigid as statuary for many seconds together. Yawning and lolling are common among tired children, and twitchings and grimaces, which in serious cases culminate in St. Vitus' dance. Here are some extracts from the various replies.

1. Sudden muscular movements. 2. Grimaces, frowning, or compression of the lips are marked signs. 3. The fingers sometimes twitch and the whole nervous system seems affected. 4. Twitching of the face. 5. Twitching, blinking the eyes. 6. Fluttering of the eyelids. 7. Tendency to nervous laughter or movements. One correspondent has fits of sneezing in the early morning when he has been fatigued over night. General unsteadiness of muscular co-ordination is shown by bad and shaky handwriting: this is sometimes specifically mentioned, but more often implied by such phrases as—8. Careless writing; or, 9. "Failure in all work requiring neatness." 10. Sometimes a loss of power to continue writing,

the pen going crooked, &c. Fatigue is also very frequently indicated by disordered utterance as—11. Tendency to stumble over words when speaking. 12. Refusal of the tongue to obey the will, so that in speaking or reading I substitute one word for another.

Irregularity of nervous action is further shown by conditions of pallor or of flushings in the face. They sometimes alternate; testifying to a depression of general nerve power, combined with morbid excitability. Allusions to abnormal skin colour are frequent in the replies. One teacher goes so far as to lay particular stress on the colour of the tips of the ears in deciding whether and in what way the girls of her class are suffering. If the tips are white, flaccid, and drooping she concludes the girls are thoroughly weary in mind. If they are relaxed but purplish, she concludes that they are "tired not with study but from struggling with their nerves, which the average school girl of 14 or 15 very rarely has completely under control."

Headaches.—The frequent occurrence of headaches in varied forms and in every degree of severity may be accepted as a matter of course. Similarly as regards cold feet, faintness and actual faintings. Sleeplessness in a very serious degree is another well-known sign; much more rarely somnolence. Grinding the teeth at night and talking in the sleep are frequently mentioned; somnambulism occasionally so. I do not propose to enter into details respecting any of the matters just mentioned, as they are all of them well known signs of over fatigue. It may, however, perhaps interest the meeting to see a drawing I hold in my hand made in sleep not many weeks ago, by a young friend and connection of my own, who was studying rather too hard for a Government examination. He awoke in the night, and found himself in his nightgown, sitting at his table with the gas burning and with this grotesque sketch of an elephant's head and of some other animals just completed. The ink was still wet. He had not the slightest recollection of anything previous to the act of awakening, but there had been conversation before he went to bed that probably suggested the sketch.

Disposition.—Irritability is perhaps the commonest sign of incipient fatigue. My correspondents freely acknowledge it to be so with themselves and it is very easily noticed among their pupils, who become cross and peevish when tired. I shall not enter further into this, as the fact is a familiar one; it is also well-known that the nerves of sensitive people becomes [sic] so irritable by overwork as to be painfully jarred by what they wholly disregard when well, such as the ticking of clocks and the rattle of the street. A most pitiable amount of suffering is disclosed in these replies, due to nervous irritability. Much is said of the gloomy way of looking at life, that is brought on by overwork; of the sense of incapacity, of magnifying trifles, and of dread of society. Irritability is sometimes accompanied by a notable amount of ordinary excitability expressed by such remarks as—1. I get nervous and start at noises. 2. I start sometimes at a sudden noise or movement in the room.

It is, I need hardly say, known by experiment, that both the quickness and the magnitude of the reaction to any stimulus is greatly affected by fatigue. There is an experiment, not so well known as it should be, that after a class had practice in performing it, can be repeated at any time in a few seconds, which gives an excellent measure of the varying amount of reaction time. The class take hands all round, the teacher being included in the circle, a watch with a seconds hand lies on the table before him. All the pupils shut their eyes. When the seconds hand of the watch comes over a division the teacher gives a squeeze with his left hand to the right hand of the pupil next to him. That pupil forthwith with his left hand squeezes the right hand of the next pupil, and so on. Thus the squeeze travels round the class and is finally received by the right hand of the teacher, who then records the elapsed time since he started it; or he may let it make many circuits before he does so. This interval divided by the numbers of pupils in the class and by the number of circuits, gives the average reaction time of each pupil. The squeeze takes usually about a second of time to pass through each dozen or fifteen persons. We should expect to find uniformity in successive experiments when the pupils are fresh; irregularity and prevalent delay when they are tired. I wish that teachers would often try this simple, amusing, and attractive experiment, and when they have assured themselves that their class enters into its performance with interest and curiosity, they might begin to make careful records at different periods of the day and see whether it admits of being used as a test of incipient fatigue. I should be exceedingly glad to receive accounts of their experiences. Deception must of course be guarded against.

Senses.—The frequency with which serious alteration in the power of hearing and of seeing is mentioned, and the feelings sometimes of intense sensitivity and sometimes of numbness, show that the delicacy of the senses is markedly affected by fatigue.

Hearing is often heightened in keenness, sometimes it is dulled. It is heightened in those numerous cases of irritability of which I have spoken, when the tired brain becomes almost maddened by an organ grinder. It is temporarily paralysed in others. The following is a mixed case:—1. My hearing had never been very acute, and I think the first symptom of fatigue is a feeling of deafness, but at the same time that I cannot hear the voices I want to hear, the outside noises of traffic, bells, &c., become intolerable. Other cases of deafness from fatigue are—2. Inability to hear in school without a painful effort. 3. Increased deafness.

Vision is greatly affected by over fatigue, not only owing to the strain upon the eyes from much reading in a bad light, but apparently through more deeply-seated causes as well. It is difficult otherwise to account for the following interesting case in which colour-blindness was brought on by fatigue and disappeared after rest. It has much physiological interest and well deserves being placed on record. The lady allows me to mention her name for the sake of authenticity. She is Miss J. Beckett, Girls' Grammar School, Ripon.

"After several hard hours of continuous study I have been subject to attacks of colour-blindness, which leave me after resting. The first time I noticed that I was not able to distinguish one colour from another, was when I was reading for the London Matriculation years ago. I was at the same time etching for an American magazine and teaching most of the day. This lasted from Christmas to July, when I began to feel considerably worn out. One day I went to spend a few hours with a friend, and whilst there, began to paint some ivy leaves on a terra-cotta plaque. Imagine my distress when my friend told me the leaves were orange instead of green. On my return I went into my study and to my astonishment, the curtains which were blue in colour looked a kind of dingy yellow. However in a few hours I was quite well. Towards the end of the year I was obliged to give up work on account of my health. I got well, and took up my work again, still subject to temporary colour-blindness when tired." In answer to further inquiries, she adds: "I do not remember whether I have any difficulty in recalling colours when tired. From a little child I have been particularly fond of them, and can readily paint flowers, foliage, and neutral tints from memory."

The frequency and severity with which the sight is affected by fatigue is sufficiently shown by the following extracts:—

1. The eyes fail first. Sometimes after hurrying to a lesson, on my arrival I could not see a single note on the page of music for a few minutes. After writing and playing long, everything goes black or black spots dance up and down. 2. A time of great excitement or worry will so affect my sight that for about half-an-hour at a time, I can see nothing clearly. The outline of everything is deficient in some part, so that I only see half of a thing at a time. There seems to be a bright wheel of light whizzing in the comer of one or the other eye. 3. At first the lines of the page become indistinct, then at intervals they appear to vibrate; finally they merge into one mass. 4. The words appear to rise from the paper and frequently a double row of words are visible. 5. Lights and after images are distinct before my eye. 6. A confusion in the lettering of mathematical diagrams is sometimes an early symptom of fatigue among my pupils. As regards sensations in the eye itself, beside such remarks as—7. A dazzling and burning sensation in the eye, the following is a case of an affection of the eye being subordinate to that of the brain, rather than vice versa. 8. A nervous sensation in the eyes as though the eyeballs were loose in my head, and would fall which ever way the head is inclined. The sensation is worse on lying down. I am somewhat short-sighted and wear glasses, but only feel this disagreeable sensation when mentally weary, not necessarily through over-reading.

Memory.—A very common and early symptom of fatigue is failure of memory, using that word in the allied senses of recalling ideas at will, or else of former ideas presenting themselves readily by association, or else of the sure association of muscular movements engaged in utterance, with the idea of the words intended to be uttered. I have made extracts of no less than twenty-five cases of failure of memory, out of which I will select half-a-dozen. 1. My first indication of mental failure is an inability to spell common words; my second, an omission of words in writing; my third, sudden forgetfulness of what I am actually saying. 2. Tendency to forget the meaning of words in a foreign language which are usually well known or have been met with quite recently. Tendency to make stupid blunders in subjects in which, when the mind is in full vigour, it is accurate without effort. Simple and obvious mistakes are increased twofold in number, and that throughout the class. 3. Through days and weeks together, the utterance of wrong words or sentences, not intended or desired to be spoken, and the writing of wrong words. 4. Tendency to stumble over words in speaking, and to misplace letters in writing, generally putting them too soon as "Wesday" for "Wednesday." 5. Want of power of calling at will to memory, names and little matters connected with every day life. 6. Some of the pupils never spell correctly when tired.

Arithmetic and Mathematics.—The studies that are the first to fail under fatigue differ in different individuals, but in the majority of cases those of arithmetic and elementary mathematics go soonest. Though many of the 116 replies come from teachers who have little, if anything, to do with those subjects, no less than forty-seven specifically mention them. For example:—

1. The merely mechanical processes of arithmetic become bewildering at the end of a day in which I have been particularly engrossed with school work.

2. Arithmetic and algebra become impossible when fatigued not as being disagreeable or painful, but because I then blunder so much that it is hardly any use attempting them.

3. Another correspondent speaks of the impossibility when fatigued of doing work that requires both accuracy of detail and a certain force of will to fix the attention, such as arithmetic.

4. Speaks of the difficulty to tired boys of working out any common sense problem in arithmetic. Though very many similar answers could be quoted in corroboration of these, there are two that tell in an opposite direction. They are—

5. Whenever my mind is wearied, it affords me a certain amount of relief to do some work which involves the solving of arithmetical and algebraical problems, and by preference such as call for the use of logarithms or of the slide rule.

6. I find accounts a great rest when I cannot exert my mind usefully in any other way. I may be permitted again to break my rule by adding a case from my own knowledge of a very distinguished man, now deceased, who, having always found repose in his favourite mathematics when he was fagged and worried by multifarious duties, naively recommended the same remedy to a friend whose brain had so broken down for a time, that he shrank from the least mental exertion as from a fatal danger.

Languages.—A difficulty in translating is another of the noticeable effects of incipient fatigue, and is partly due to the lapses of memory already spoken of.

1. In translating, words and phrases do not occur readily to the mind.

2. Translation into or out of a foreign language with which I am not very familiar.

3. I have occasionally lost the power of speaking German when fatigued, though when in my ordinary condition I speak it without conscious effort.

The failure to translate well is due of course to much more than the simple failure of memory in small things and depends on the loss of power of grasp, and on depressed mental vigour generally. The following is an instructive case:—

When I taught young boys of ages 8 to 13, all day, I took arithmetic and Latin in the morning, and English reading, geography, &c., in the afternoon. On some occasions the Latin lesson got put off till the afternoon, and I was surprised to find that lesson, which was always a successful one in the morning, failed entirely in the afternoon. The boys wished to learn but could not. Their ordinary work, which made less demand on the intellect, they did in the afternoon well enough. This and such like cases fall more properly in the next division.

Failure of Mental Grasp.—The evidences that the fatigued mind is unable to work up to its normal standard, and that it wastes itself in futile exertion, are very numerous. They are such as:—

1. Failure of ability to grasp the meaning of even simple things.

2. Failure of the portative memory. In reading complete inability to take in the matter whilst mechanically scanning the page. A curious incapacity to count the cups when serving tea.

3. Reading sentences without recognition of what was read.

4. Confusion alternating with excessive clearness of thought.

5. Tendency of thoughts to wander. Failure in pupils to grasp the meaning of what is said to them quickly and fully.

6. Before the actual sense of fatigue is distinctly felt, I am conscious of a want of power to grasp ideas, and of an incapacity for conveying them clearly.

7. Inability to read the 'Journal of Education.'

8. Rapid disappearance of immediately preceding concepts, and hence difficulty in establishing connections between paragraphs, as in writing a Review article.

9. Tendency to use long words. (This strikes me as a very suggestive reply).

10. Any book in which the language is wanting in ease and simplicity, though its subject may be familiar or easily understood. In short, to use a common and vigorous phrase, the mind ceases to bite, when it is fatigued.

Failure of Energy.—It requires no evidence to corroborate the well known fact that energy fails as fatigue increases. New subjects are distasteful; teaching dullards becomes almost an impossibility. Sustained effort, vigorous inspection, quick decision—all are impossible.

Possibility of Tests of Incipient Fatigue.—The replies I have received do not contain any distinct proposition of tests of incipient mental fatigue, and I am myself far too ignorant of the practice of education to venture to formulate any. On the other hand, the replies are not deficient in indications of what such tests might be directed to ascertain. They are principally as follows:—

1. The length of time during which neatness of execution can be sustained in performing a prolonged task.
2. Promptness and sureness of memory in simple things.
3. Common sense arithmetical problems.
4. Reaction time.

The measure of fatigue is inversely the measure of endurance, and this strikes me as being a faculty that well deserves investigation. Under the strain and exhausting calls of modern civilized life, the power of endurance is rising continually in importance. Men and women have now a-days to act rapidly and for many hours, and not only to act exceptionally well. It therefore seems very reasonable that teachers should direct their attention to some fair way of appraising the relative power of endurance among their pupils. It is of course incidentally discovered in the ordinary course of tuition, but one would like to see appropriate tests directly applied to determine it, and such as would show at any time in a definite and unmistakeable manner whether the minds of pupils were fagged or not.

Breaking Down.—I now come to the evidence given in these replies respecting the frequency with which both pupils and teachers are found to "break down." There is an intelligible and very transparent tendency in not a few of the respondents to say that such a thing as overwork is impossible in their respective schools. Some of them protest so much and so extravagantly as to raise not a little suspicion. There are even a few who say they have never heard of a case of breaking down.

Taking all the replies together, I find that out of my 116 correspondents no less than 23 of them have at some period of their lives broken down, and that 21 of these have never wholly recovered the effects. There are six other cases of a less serious kind, some of them slight. In other words one out of five teachers has, so far as the evidence before me goes, been severely stricken.

As to the cases well known to my correspondents, there is vagueness in some of the replies where the word "several," and the like, are used, to which I am quite unable to assign a numerical value, but 59 sad cases are specified in detail in answer to the question 5: "Have you known cases of more or less serious prostration from mental overwork, as distinguished from the effects of domestic or other anxiety? If so, give initials and dates, and a very brief notice of the severity and duration of the illness." In many other cases the writers express the difficulty they feel in distinguishing between worry and overwork. The latter is a consequence of the former, while the former often results from the gloom, anxiety, and sense of incapacity caused by the latter. It is a self regenerating circle of evil.

I draw two conclusions from the replies. The first is that the reason why mental fatigue leaves effects that are so much more serious than those of bodily fatigue is largely owing to the cause just mentioned. When a man is fatigued in body he has very similar symptoms to many of those mentioned above, but there is a great after difference. As soon as the bodily exertion has closed for the day, the man lies down and his muscles have rest; but when the mentally fatigued man lies down, his enemy continues to harass him during his weary hours of sleeplessness. He cannot quiet his thoughts and he wastes himself in a futile way.

The other conclusion is that breaks down usually occur amongst those who work by themselves, and not among pupils whose teachers keep a reasonable oversight. Over zealous pupils are rare, as many of my correspondents insist. But the danger is not so much at school, when the hours of study and those of play and exercise are fixed, as it is at the age when young persons are qualifying themselves for the profession of a teacher, and who have also to support themselves, and perhaps to endure domestic trials at the same time. Dull persons protect their own health of brain by refusing to overwork. It is among those who are zealous and eager, who have high aims and ideas, who know themselves to be mentally gifted, and are too generous to think much of their own health, that the most frequent victims of overwork are chiefly found.

DISCUSSION.

Mr. J. G. FITCH, Her Majesty's Inspector of Training Colleges,[3] remarked that the returns collected with so much care by Mr. Galton were very interesting and suggestive, and might, if extended over a wider area, prove very useful. Meanwhile it should be remembered that their trustworthiness and value depended, in some measure, upon conditions which had not been referred to in the paper, and on which no information was before the meeting. In judging of the signs of incipient weariness or over-work in a class, the one consideration of most importance as a factor in the problem was the character of the teacher and of his lesson. A bright earnest teacher, who knew how to kindle the interest and sympathy of his scholars, would observe few or no tokens of fatigue, while a dull, spiritless and mechanical teacher might find his pupils restless or yawning before he had been five minutes in their presence. If there were two teachers, the one of whom was gifted with real aptitude and power and knowledge of method, and the other of whom was deficient in these qualifications, both might be at work in classes of the same number, the same age, and the same capacity, and might be teaching the same subject; yet their reports in answer to Mr. Galton's questions would differ substantially; for they would not only observe in a different spirit, but they would have very different phenomena

[3] Joshua Gurling Fitch (1824–1903).

to observe. Unless we know something of the character and capacity of the teachers, it would be unsafe to deduce any general inferences from their testimony on the mental fatigue of their scholars. Another condition very materially affecting the whole problem was the character of the time-table in use in the school. If the subjects of instruction were wisely distributed; if the lessons were not of undue length; if the various occupations of the day were so varied that light mechanical exercises alternated with lessons which required serious mental application, there need be little or none of the irritability or the languor so often complained of. Whether school-work was wearisome and hurtful or not depended not so much on its amount as upon the skill and good sense with which the work was planned, and with which the different mental and bodily faculties were called into exercise in turn. He did not say this to detract from the value of the very interesting enquiries which had been described in Mr. Galton's memoir, but merely to suggest caution in deducing inferences from the answers unless more was known about two of the most prominent and significant of the conditions of school-life—the quality of the teaching and the distribution of the employments.

Mr. STRAKER,[4] following the suggestion of the previous speaker, that the number of subjects probably influenced the amount of mental fatigue, said that he had worked at three subjects, taking them in turns, for two years, and that he was unable to do any work for about a year afterwards, from the resulting mental fatigue. The work was almost continuous; the subjects were absolutely unconnected with each other; two of them required much thought, one of the two a good deal of writing; the other was ordinary preparation for an examination, and was the easiest of the three, though taking about an equal amount of time. As to the amount of work, any two of the subjects would have been full ordinary employment. Therefore intervals of time which would have been occupied in relaxation were devoted to work. He had previously suffered from mental fatigue, for a few weeks after his first examination at Cambridge, which he had prepared for in a very short time; also for many months after his Tripos, in which he took a low place in honours.

---

[4] Joseph Straker (1858–1930), a barrister.

Chapter 23

How the Anthropometric Laboratory Grew

Dec 31 / 1893[1]

It may be well to explain how the Anthropometric Laboratory & Finger prints grew out of previous works. After writing Hereditary Genius and other kindred subjects, I tried to measure heredity, whence my lecture on peas (Typical Laws of Heredity) at Royal Institute in 1877. This gave a law, but I did not see my way to verify it on men, so I took up various faculties, trying to find suitable ones to measure. Hence the book Human Faculty. This led me to Psycho-Physics and to the idea of getting up a laboratory for experiments, whence frequent relations between myself and Croom Robertson. Then, the International Health Exhibition of 1884 coming in, I proposed to start a laboratory there, which was a success. I had also offered prizes for family records, the governing idea in both cases was to lay the foundation of ways of recording measurable data that hereafter might serve for strict hereditary conclusions. It turned out that the family records, lax as they were in many respects, did afford data of the kind I wanted, whence I worked out the law of regression which entirely confirmed what the peas had shown. This was the subject of my address at Aberdeen to Section H in 1885, and of the book "Natural Inheritance". Then, with Mr Merrifield's help, I tried moth breeding (selenia illustraria[2]) 2 generations a year for 4 years—A, B & C hereditary sets, but the broods were not large enough for statistics & the thing, which had given much trouble, failed. In 1888 Bertillon's system was

1 GALTON/2/4/18/3.

2 Also known as Selenia tetralunaria.

being carried on in France & I gave a lecture at the R. Inst. on Personal Identification, in which I spoke of finger marks, having stumbled in "Nature" on a short account of Sir W. Herschel's work in Bengal, which I inquired about. At this time too, I had started my laboratory in S. Kensington. I had wanted among other things to introduce marks for physical efficiency in competitive army &c. examinations & the laboratory helped in getting data for this. Finger marks fascinated me on account of their unique persistence, & the belief that they might be (which they are not) of ethnological &c. value. Being anxious to get to the bottom of this & to use their peculiarities to the uttermost, I have given perhaps an undue amount of attention to them. However it seems, after all, likely to lead to useful practical results.

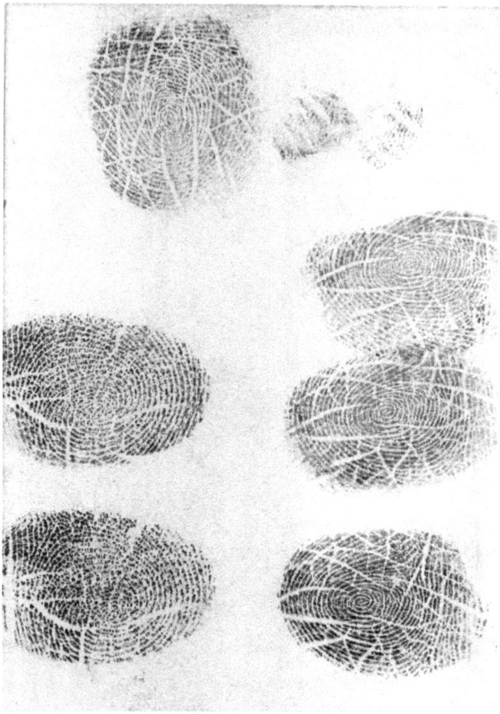

Figure 23.1: Francis Galton's Fingerprints

Chapter 24

Souvenirs d'Egypte, 1900

Francis Galton
President of the Anthropological Institute of London.[1]

Excellencies, Ladies and Gentlemen,

Allow me, above all, to express to you the pleasure and the honor that I feel in being in the midst of a group of representatives of geography, in a country such as Egypt, unique in the world by its situation and by the extraordinary antiquity of its historical monuments.

When you asked me, Mr. President, to make a presentation at this meeting, my first idea was to deal with the differences that struck me the most between Egypt today and the way it was when I visited it in 1846, fifty-four years ago.

The changes are really substantial. At that time the crocodiles were basking on the sand-banks and positively dangerous in Assiout;[2] there was a greater quantity and a greater variety of birds. In addition, enough time has elapsed for a noticeable change in the general level of the surface of the alluvium, which amounts, it is said, to annual deposits in the proportion of one meter per thousand years. In this case the ground would have increased by 34 millimetres: it would be two inches higher than during my last stay here. In other words, the monuments would have sunk by so much.

But the state of Egypt in 1846 and even before can be better seen by the consultation of works and documents published on this subject, than by the summary remarks that I can present to you today. I prefer to seize this opportunity to

[1] Galton 1901. Translated from the French by GT. See report of the meeting of February 24, 1900.

[2] Assiut, or Asyut.

offer a tribute of gratitude to your eminent colleague, the late D'Arnaud Bey, whose portrait in the gallery of the scientists and explorers of your museum I have seen with great interest and emotion. It was actually at his instigation that my trip to Khartoum took place in 1846, a trip which had such a great influence on my future career and which decided my vocation. Here's how:

Two friends (who have been dead for a long time) and I were sailing as tourists to Wadi-Halfa, when the march was halted by the contrary winds at Korosko. D'Arnaud Bey was there, returning from a scientific mission to the Sudan, which had been entrusted to him by Mohamed Ali. We made acquaintance. He lived in a little mud hut, surrounded by a slight palisade of reeds, which his genius had transformed into a philosopher's sanctuary of sorts, in the midst of an almost barbarous country. The artistic arrangement of the small collection of scientific instruments, books and curiosities, gave a singularly tranquil and studious air to that modest dwelling. I was delighted by his conversation and the many stories he told us about the countries of the South. He told us, "Why do you just hang out as simple tourists? The road to Khartoum is open and a sheik with his camels is here. Make your conditions with him and send your boat to Wadi-Halfa, where you can take it back on another way."

We accepted the proposal with enthusiasm and left three days later, with a rudimentary map as a guide. It is not necessary to dwell on the details of this excursion; the places we visited are now familiar and accessible; but then it was not so. We went through the desert, first to Abu Hammed, then to Berber, from where a chartered boat took us to Khartoum. After going up the White Nile, we went down again to Metemna. From that place we crossed the desert of Bajuda to Merawi by camel, and by way of Dongola we reached Wadi-Halfa.

I did not have the opportunity, since then, to meet up with or get in touch with Arnaud Bey, but I have always considered him in my heart as a benefactor, because it is he who gave me the idea of the purposeful journeys I made later exploring the land of the Damara in South-West Africa; it was he who directed my taste and energy towards geography and science; so that I knew how to use my time usefully instead of devoting it to frivolous amusements.

Another fact, of which I have recently become acquainted,

also draws me to the memory of D'Arnaud Bey; this is a recent application of some of my anthropological research. In a certain sense, it is still Arnaud Bey who prodded me onto this path, awakening in me the love of scientific research.

A few years ago I began studying fingerprints as an anthropological character. I began by noticing and establishing, as a fact beyond doubt, that the faint furrows which form such singular patterns at the end of the fingers have an indisputable permanence. So I spent a lot of time collecting and sorting these drawings in the hope of drawing distinct signs of race, or rules used to mark hereditary laws. Unfortunately this hope was more or less disappointed, but my research gave me positive results from another point of view.

The certainty that I had acquired of the permanence not only of generic imprints, but of the furrows that compose them, in their most minute peculiarities, and which can be compared to the plan of a city, encouraged the view that fingerprints should be considered as an irrefutable mark for the identification of criminals, and that, in view of the experience I had in classifying subjects, it might be that a dictionary of these marks, somehow lexiconized, if you will allow me the word, would facilitate research on identification.

In short, I demonstrated that fingerprints can be used according to the same principles as the measurement of the head and other limbs, according to A. Bertillon's method, adopted by the Identification Office in France. I submitted my ideas to the public in various forms, and it was then that the British Government appointed an Interim Commission in order to draw up a report on the comparison of the two systems, that of Mr. Bertillon and mine, and on their application, whole or partial, in England. The report advocated a combination of the two systems applied with caution and moderation, with which I fully agreed. The different dimensions of the same individual are so correlative to each other that it is useless to employ many, but they are not correlative, up to a certain point, with the imprints of the different fingers. Consequently, the combined use of the two means of observation gives a more perfect assurance of the identification of an individual.

A combined system of measurements and fingerprints is now in use in England. Thanks to the energetic initiative of Colonel Harvey Pasha, this system is also in use in Cairo; the latter was kind enough to show me his system at work, and I am happy to report that the rapidity and regularity of the observations are worthy of all praise. In the four or five cases I saw in practice, research to find the original class lasted less than a minute for each case. It is more than likely that the gradual extension of the study of fingerprints will lead to results other than the goal of identifying a criminal. Indeed, in India, for social reasons we had to dismiss the measurements, which is very regrettable, but the imprint has been used in several civil cases (nine or ten). But I must not let myself be distracted by a subject that would lead me very far astray. I wanted to mention it to declare that the results that I obtained in this area of study are ultimately due to the geographical enthusiasm that was awakened in me by the lamented Arnaud Bey, in the few hours that I had the pleasure of passing with him, in 1846, in Korosko.

Chapter 25

Our national physique—prospects of the British race—are we degenerating?

The alleged deterioration of our race through town life and other insanitary conditions is indeed worthy of serious inquiry by a Royal Commission or otherwise, but the points which a commission is likely to consider can only be part of a much larger object that well deserves to be kept in evidence before the public.[1] A strong consensus of educated opinion in this larger sense is needed to stimulate the inquirers to their utmost, and to prepare the public to accept the measures that the Commissioners may eventually think fit to propose, even though they be unexpectedly drastic. In the following remarks I propose to deal with the subject on these wider lines, and shall not confine myself to what relates only to the wastrels and to the slums. The questions I keep before me are whether or no the British race as a whole is, or is not, equal to its Imperial responsibilities, and again, how far is it feasible to make it more capable of the high destinies that are within its reach, if it possesses the will and power to pursue them. I wish that each one of us should stand aloof from ourselves as a whole, and should watch the conditions and doings of our race, much as an authority of the Royal Agricultural Society might criticise the stock of his neighbour over the hedge. If we do so we may learn in what ways our own stock and its rearing are open to improvement and we may perhaps [ensure][2] it.

There is no question that the pick of the British race are as

1 Galton 1903.
2 'Ensue' was printed but that appears to be a typo.

capable human animals as the world can at present produce. Their defects lie chiefly in the graceful and sympathetic sides of their nature, but they are strong in mind and body, truthful and purposive, excellent leaders of the people of lower races. I speak more particularly of those who are selected to go abroad in various high capacities, whether by Government or by firms to carry out large undertakings under circumstances where they have to depend much on themselves. The lower middle-class of Britons are quite as efficient by nature, probably they are more so than men of other races, but they are of a coarser fibre than the Latins. Our average holiday-maker and cheap-excursion tourist is proverbially unprepossessing. He probably admits of high polish, but the process is far too costly for his means, so he stands on a conspicuously lower social level than those of his compatriots who have been more favoured in their fortunes. His manners are noisy and vulgar, "they have not the repose that marks the caste of Vere de Vere," which men of other races have. As regards the physique of Britons, I think we brag or have bragged more than is right. Moreover we are not as well formed as might be. It is difficult to get opportunities of studying the nude figures of our countrymen in mass, but I have often watched crowds bathe, as in the Serpentine, with a critical eye, and have always come to the conclusion that they were less shapely than many of the dark-coloured people whom I have seen.

There was a curious exhibition not very long ago at the Albert Hall of men selected for strength, which, so far as I saw, attracted less attention from the newspapers than it deserved, so I will mention my own impressions. If I am inaccurate through writing from memory alone the faults will, I am sure, be in the details of secondary importance. Sandow, the well-known athlete, has established schools of instruction in places variously scattered over England, and he arranged two or three years ago to give no less than £1000 in prizes to the three best-made men among the numerous pupils in his many classes.[3] During a tour of inspection he selected from each class the three whom he considered the best, and in due time the whole body of selected competitors were assembled in Albert Hall at a public meeting. They were dressed alike in tight trousers, which showed the shape, and each had a

[3] Eugen Sandow (1867–1925) the Königsberg bodybuilder. Galton is describing the world's first bodybuilding competition, which took place in the Royal Albert Hall on 1901/09/14.

scarf thrown jauntily over his otherwise nude shoulders. Ten solid pedestals stood in front of the organ; electric light was in readiness, and a brass band occupied a corner of the orchestra. I will not dwell on the many and very prettily arranged gymnastic feats that were shown, but shall speak only of what is now to the point. The candidates, about eighty in number, marched in a phalanx, ten men abreast.[4] The first row detached itself, went nimbly to the ten pedestals, where each man threw off his scarf and mounted upon one of them. The electric light was flashed on the ten standing figures. Three judges (of whom Sir Conan Doyle was one[5]) walked round them, note-books in hand, criticising and appraising. Sandow was in readiness to act as referee if called upon. Then the best man of the ten was selected by the judges, and he stepped apart. The same was done to each row in succession until all were gone through. Then the twice-selected competitors (I think there were eight of them[6]) mounted the pedestals side by side. They were scrutinised even more closely than before just as prize cattle. At length the final selections were made, and the victor was handed down the Hall, to the tune of "See the conquering hero" from the brass band, and received his magnificent award of a gold trophy amidst the plaudits of all; the second similarly received a silver replica of it, and the third a bronze one. I studied the whole of the proceedings carefully through an opera glass, and had by my own judgment fixed upon two of the three during the preliminary stages. I should say that the first of all had not been one of these, but I fully recognise the justice of that award in the end.[7] Now as to my conclusions. I did not think those best specimens of the British race to be ideally well-made men. They did not bear comparison with Greek statues of Hercules and of other athletes, being somewhat ill-proportioned and too heavily built. I must say that I was disappointed with them from the aesthetic point of view, though in respect to muscular power they seemed prodigies. Sandow afterwards exhibited himself in a pose that brought out his chest and arms to full advantage,

---

4 There were 60 competitors.

5 Sir Arthur Conan Doyle (1859–1930), author of the Sherlock Holmes stories and other works, was one of Galton's friends. Sir Charles Lawes (1843–1911), a sculptor and oarsman, was the other judge.

6 There were twelve.

7 In his memoirs, Galton added that 'The power of classifying men correctly, by mere inspection, seemed to me much greater after this experience than before', Galton 1908b, 280.

and in that statuesque position I placed him as far superior to all the competitors.

I therefore doubt if a stock breeder could make very much out of the British race in respect to physical beauty of the trunk and limbs. Subject to this small reservation, the imaginary critic above mentioned could emphatically affirm with justice that the whole of a race which was able to furnish the large supply that is produced in Great Britain of men who are sound in body, capable in mind, energetic and of high character, has the capacity (speaking as a rearer of stock) of being raised to at least that same high level. How to do this is a question of both Nature and Nurture. I shall not reiterate here what I have on more than one occasion urged in respect to the former of these two great influences, further than to repeat my conclusions, which were, first, that if a strong and intelligent public opinion should ever become aroused in favour of improving the breed of our race, influences are in existence which even now operate in a small way in entire conformity with sentiment and law, and which are capable by development of producing great results. Individually their power is small, but in cooperation they would wield a practically overwhelming force. The second conclusion was that we have a great deal yet to learn on matters bearing upon race-improvement, which lies within the plain province of anthropology, before it would be justifiable to attempt a crusade. Otherwise great mistakes would arise, efforts would be wasted, the results would not agree with anticipations, and the movement would become discredited and collapse. My attitude, which has usually been misrepresented, is to urge serious inquiry into specific matters which still require investigation in the well-justified hope that a material improvement in our British breed is not so Utopian an object as it may seem, but is probably quite feasible under the conditions I have just named. But whatever agencies may be brought to bear on the improvement of the British stock, whether it be in its Nature or in its Nurture, they will be costly, and it cannot be too strongly hammered into popular recognition that a well-developed human being, capable in body and mind, is an expensive animal to rear. It would be bad economy to spend more on him than can subsequently be got back while it is equally bad economy to starve him into weakness. A horse must be well fed in order that the value of his work shall bear the highest possible ratio to the cost of his food. It was remarked by an eminent contractor that in all lands the wage and the amount of work done for it bore a curiously constant proportion, low wages producing little, high wages much. It is, of course, a matter of careful inquiry to learn the limits within which this alleged law holds good. The most appropriate amount of feed for horses is known well enough.

When considering how to deal most economically with the masses, a question arises which seems to me of much importance, but which has never to my knowledge been put, and which I trust that the Commission will entertain. It may be phrased in this form. A certain small dole is available for each child to help his rearing; at what age would that exceptional and temporary assistance be most effective? Is it in childhood, or at about the time of most rapid growth? The immense improvement in the physique of previously ill-fed recruits, after a year's good feeding with an out-of-doors life, is well known, but much irremediable mischief may have been done before that age is reached. Would it be better to begin earlier and to leave off sooner? In short, a careful inquiry should be made as to the maximum of good effect on the physique that might on the average be produced through the extra expenditure of a given sum of moderate amount, paying especial regard to the ages at which that sum would have to be expended.

Figure 25.1: Francis Galton in 1909 at age 87, with his manservant Rudolf Albert Gifi.

Chapter 26

Address on Eugenics[1]

[This address was delivered by Mr. Francis Galton, F.R.S. (author "Hereditary Genius; Inquiries into Human Faculty," and several other works on Anthropology), in compliance with a request of Mr. Crackanthorpe,[2] at 65, Rutland-gate, on June 25, 1908.]

The Science of Eugenics (pronounced with a soft g, as in the name of the Empress Eugenie and in the word Genesis) is based on Heredity. Permit me to begin my remarks by a few words on this subject, and on what first led me to study and write about it, more than forty years ago.

It may surprise some of you to learn that the very word "heredity" was not then in common use; it does not appear in Johnson's Dictionary, and, though it has since become familiar, I was chaffed by a cultured friend for adopting it from the French.

The notions connected with human inheritance by descent were vague and confused, for the subject had never been squarely faced. The prevalent feeling was that it certainly existed in animals and plants, but that men stood in a separate category. It was acknowledged that physical attributes were sometimes inherited in human families, but the heredity of mental qualities in man was stoutly denied by many, especially by theologians on purely dogmatic grounds. There was much talk about men being equal and "masters of their own fate."

I date my first opinion from my days at college, as far back as 1840, where competitions of all kinds showed most clearly to an unprejudiced eye that men are not equal in their natural powers, but most diverse in mind as well as body. It was also

1 Galton 1908a.
2 For Crackanthorpe, Montague, see page 355.

noticeable that high gifts of both of these tended to run in families.

I think the first evidence of the kind that strongly impressed me was in relation to classical successes. To be a Senior Classic—that is, to be the very first classic of the year at Cambridge, where the body of undergraduates contains picked boys from schools in all parts of the country—is a very considerable feat. Yet I found that these Senior Classics were often so closely interrelated that out of forty-one of them six had either a father, son, or brother who was a Senior Classic, or in one case a Senior Wrangler. No mere tuition could account for this. They must have been born with exceptional capacity.

I found the same to be true of every form of bodily and mental activity into which I inquired; those who achieved the most had a far larger proportion of eminent relatives than accident or good teaching could account for.

There were then no means such as we have now—thanks to the development of statistical science—of measuring with numerical exactness the closeness of the various kinships. That was a subsequent achievement. Neither did I then go into the question of fertility as has since been done.

It gradually became more and more clear to me that man was not such an exceptional creature after all in respect of heredity, but that what applied to other animals and plants probably applied also to him. I perceived that the importance ascribed by all intelligent farmers and gardeners to good stock might take a wider range. It is a first step with farmers and gardeners to endeavour to obtain good breeds of domestic animals and sedulously to cultivate plants, for it pays them well to do so. All serious inquirers into heredity now know that qualities gained by good nourishment and by good education never descend by inheritance, but perish with the individual, whilst inborn qualities are transmitted. It is therefore a waste of labour to try so to improve a poor stock by careful feeding or careful gardening as to place it on a level with good stock.

The question was then forced upon me—Could not the race of men be similarly improved? Could not the undesirables be got rid of, and the desirables multiplied? Evidently the methods used in animal breeding were quite inappropriate to human society, but were there no gentler ways of obtaining the same end, it might be more slowly, but almost as surely? The answer to these questions was a decided "Yes," and in this way I lighted on what is now known as "Eugenics."

Eugenics has been defined as "The study of those agencies which under social control may improve or impair the racial qualities of future generations, either physically or mentally." It aims at showing clearly how much harm is being done by some one course of action, and how much good by some other, and how closely connected social practices are with the future vigour of the nation. Its procedure is the reverse of fanatical; it puts social problems in a clear white light, neither exaggerating nor underrating the effects of the influences concerned. It is probable that even democratic governments will hereafter appreciate the value of Eugenic studies, and deduce from their results recognised guides to conduct. Such governments would be compelled to do so in their own self-defence, if not on higher grounds; otherwise they would come to an end, for a democracy cannot endure unless it be composed of capable citizens.

The influence of public opinion, together with such reasonable public and private help as public opinion may approve of and support, is quite powerful enough to produce a large, though gentle, Eugenic effect. It is already becoming possible through Eugenic study to foresee with much assurance that such-and-such proposed action will influence a definite percentage of the population, though we cannot at present, and probably never shall be able to, foretell whether the individuals so affected will be A, B, C, or X, Y.

To the statesman this individualisation is unimportant, since individuals are only pawns in the great game which he plays. The true philanthropist, however, concerns himself both with society as a whole and with as many of the individuals that compose it as the range of his affections is wide enough to include. If a man devotes himself solely to the good of the nation as a whole, his tastes must be impersonal and his conclusions appear to a great degree heartless, deserving the ill title of "dismal" with which Carlyle labelled Political Economy. If, on the other hand, he attends only to certain individuals in whom he happens to take an interest, he becomes guided by favouritism, oblivious alike of the rights of others and of the well-being of future generations. Statesmanship is concerned with the nation; Charity with the individual; Eugenics is concerned with and cares for both.

A considerable part of the huge stream of British charity furthers, by indirect and unsuspected ways, the production and support of the Unfit. No one can doubt the desirability of money and moral support, now often bestowed on harmful forms of charity, being directed to the opposite result, namely, to the production and well-being of the Fit. For the purpose of illustration we may divide newly married couples into three classes according to the probable civic worth of their offspring. Amongst such offspring there would be a small class of "desirables," a large class of "passables," and a small class of "undesirables." It would surely be advantageous to the country if social and moral support, as well as timely material help, were extended to the desirables, and not monopolised, as it is now apt to be, by the undesirables.

Families which are likely to produce valuable citizens deserve at the very least the care that a gardener takes of plants of promise. They should be helped when help is needed to procure a larger measure of sanitation, of food, and of all else that falls under the comprehensive title of "Nurture" than would otherwise have been within their power. I do not, of course, propose to neglect the sick, the feeble, or the unfortunate. I would do all that available means permit for their comfort and happiness, but I would exact an equivalent for the charitable assistance they receive, namely, that by means of isolation, or some other less drastic yet adequate measure, a stop should be put to the production of families of children likely to include degenerates.

One word as to what is being done to further the prospects of Eugenics:

In the first place its study is being seriously carried on under the shelter of the University—not of Gulliver's Laputa—but of this London of ours. A Research Fellow and a Research Scholar are engaged all day on difficult statistical inquiry, rooms for the purpose being allotted to them in University College, Gower Street, under the official title of the "Eugenics Laboratory." This laboratory is now under the general care of Professor Karl Pearson, who, notwithstanding his onerous duties and pursuits, has most kindly undertaken its charge for the time, I being too old and infirm to attend properly to it. During the two and a half years of its existence it has done a large amount of substantial work. I have by me lists of five published "Memoirs," of five more which are nearly or quite completed but as yet unpublished, and there are besides six others in hand. It must be confessed that these Memoirs are by no means light reading, being crammed with figures and formulae.

Supplementary to the Eugenics Laboratory is the new and promising "Eugenics Education Society," under the presidency of Sir James Crichton-Browne,[3] the object of which is to popularise eugenic results and turn them to practical account. Young as the society is, it has already been listened to with respect by the London County Council on the question what to do with inebriate women who do not appear to be curable, and it has made useful suggestions to the Departmental Committee recently appointed by the Home Secretary to consider the amendment of the Habitual Inebriates Acts.

I will only add to this brief address that my purpose will have been fulfilled if I have succeeded in impressing on you the idea that Eugenics has a far more than Utopian interest: that it is a living and growing science, with high and practical aims. I would ask you to make the Society known to your friends, and to persuade them as best you can to help on its good work.

[3] 1848–1938. A distinguished psychiatrist, at that time a 'Lord Chancellor's Visitor' to 'chancery lunatics'.

Chapter 27

Timeline 1840–69

1840

1840/10/13.	[Tuesday. Samuel Tertius and Francis Galton leave Leamington for Cambridge on the Rising Sun coach. They stay at the Hoop Inn in Cambridge.]	Cambridge.
1840/10/14.	[Wednesday. STG and FG tour colleges and see rooms in Trinity College and inspect the Cambridge Gaol.]	Cambridge.
1840/10/15.	[Thursday. After FG takes his rooms in Trinity, STG departs for Leamington.]	Cambridge.
1840/11/18.	[Coached by O'Brien. Serious illness leading to a 12 day strict confinement to bed]	Cambridge.
1840/12/06.	[STG arrives in Cambridge to check on his son.]	Cambridge.
1840/12/12.	[FG arrives back in Leamington.]	Leamington.

1841

1841/01/04.	[In Leamington, FG sits for portrait by Oakley.]	Leamington.
1841/01/23.	[FG sees Bramah in London about his lock design.]	London.

1841/02/03.	[FG is back in Cambridge but ill again. Working on rotary steam engine.]	Cambridge.
1841/03/31.	[FG is at Claverdon.]	Claverdon.
1841/04/08.	[FG is back at Cambridge. Defers Trinity Examination.]	Cambridge.
1841/06/—.	[Maths and classics examinations.]	Cambridge.
1841/06/11.	[FG leaves Cambridge for London, where he stays at the Old Hummums hotel.]	London.
1841/06/15.	[FG arrives in Leamington.]	Leamington.
1841/06/29.	[FG leaves Leamington for Keswick via Birmingham.]	
1841/07/01.	[FG is in Keswick with the Mathison and Eddis reading party from Cambridge]	Keswick.
1841/09/15.	[STG visits FG in the Lake District.]	Keswick.
1841/10/14.	[FG leaves Leamington for Birmingham.]	Leamington.
1841/10/19.	[FG leaves Leamington for Cambridge.]	Leamington.
1841/10/20.	[FG is at Cambridge. Coached by Hopkins.]	Cambridge.
1841/12/16.	[FG arrives back from Cambridge.]	Leamington.
1841/12/23.	[STG and FG attend Lafontaine's lectures on mesmerism.]	Warwick.

1842

1842/01/08.	[FG leaves Leamington on the Eagle for Cambridge.]	Leamington.
1842/03/02.	[FG passes the Lent Term 'Little Go' ('Previous Examination').]	Cambridge.
1842/04/12.	[FG is returned to Claverdon on account of ill-health.]	Claverdon.

1842/05/—.	[FG is recuperating.]	Leamington.
1842/06/15.	[FG leaves for Aberfeldy via London to join a Cambridge reading party with Arthur Cayley and Venables.]	Leamington.
1842/06/19.	[FG is in Aberfeldy, via Dundee. Cannot study due to health, 'my head scarcely improves'.]	Aberfeldy.
1842/07/30.	[FG attends a Highland wedding.]	
1842/09/14.	[FG is at M'Queen's Hotel in Edinburgh. Queen Victoria's visit to Scotland.]	
1842/09/30.	[FG arrives in Brighton.]	Brighton.
1842/10/14.	[FG leaves Brighton for Cambridge via London.]	Brighton.
1842/11/02.	[FG is at Cambridge but his 'head' is 'uncertain'.]	Cambridge.
1842/12/23.	[FG arrives from Cambridge looking unwell.]	Leamington.

1843

1843/01/—.	[FG sees Dr. Jephson.]	Leamington.
1843/01/31.	[FG leaves Leamington for Cambridge.]	Leamington.
1843/02/16.	[Peacocke and Craufurd fight]	Cambridge.
1843/04/—.	[FG arrives in Leamington from Cambridge.]	Leamington.
1843/04/05.	Saw Dr. Jephson.	Leamington.
1843/04/12.	Went to Leamington & saw Dr. Jephson. Rx. Decoct Taraxaci 3x[1] ...	Leamington.
1843/04/17.	Saw Dr. Jephson [and Dr. Hodgson.]	Leamington.
1843/04/24.	Saw Dr. Jephson.	Leamington.

[1] Dandelion extract.

1843/05/—.	[FG is back at Trinity. Attends John Stevens Henslow's lectures on Botany.]	Cambridge.
1843/05/19.	[First meeting of the Epigram Society.]	Cambridge.
1843/06/05.	[FG leaves Cambridge for Claverdon.]	Cambridge.
1843/06/13.	[FG and EG leave London] Hamburg, Magdeburg, Berlin, Dresden.[2]	Germany.
1843/08/03.	Saxon Switzerland to 5th August.	Germany.
1843/08/16.	[FG is in Dresden.] No. 9 Johannisberg	Germany.
1843/08/17.	Left Dresden to the Noels at Felschen to August 19th	Germany
1843/08/19.	Toeplitz to Aussig	
1843/08/22.	Prague on 22nd, Carlsbad, Eger, Amberg, Ratisbon	
1843/09/02.	To Munich 22 Dultz Platz, Augsburg Lindau, Constance, Hollenthal, Kehl, Nonnenwirth, Cologne, Aachen, Ghent, Ostend	
1843/10/14.	[FG is back in Leamington dining with Dr. Jephson.]	Leamington.
1843/10/20.	[FG is in Cambridge, shortly before the Queen's visit.]	Cambridge.
1843/10/—.	[The Galtons go up to Cambridge for the Queen's visit.]	Cambridge,
1843/11/03.	[Historical Society meets for the first time, in the Black Bear.]	Cambridge.
1843/11/—.	[FG joins Silk Buckingham's British & Foreign Institute.]	Cambridge.

1844

[2] German tour details are from Emma Sophia Galton's Diary.

1844/01/20.	[FG graduates BA from Cambridge.]	Cambridge.
1844/01/24.	[FG returns home from Cambridge.]	Leamington.
1844/02/—.	[FG is back in Cambridge but ill. He attends medical lectures: Chemistry, Anatomy, Medical Jurisprudence, Material Medicine.]	Cambridge.
1844/03/—.	[STG is seriously ill.]	Cambridge.
1844/03/29.	[FG arrives home from Cambridge.]	Leamington.
1844/04/09.	[FG visits his uncle Dr. Robert Waring Darwin.]	Shrewsbury.
1844/04/20.	[FG leaves Claverdon for Cambridge.]	Claverdon.
1844/05/01.	[FG is at Cambridge, takes up abstinence from ardent spirits.]	Cambridge.
1844/05/—.	[M.P.W. Boulton and J. Kay graduate BA.]	Cambridge.
1844/06/12.	[FG is in Covent Garden.]	London.
1844/06/18.	[FG enters as a pupil at St George's Hospital. He is also registered at King's College.]	London.
1844/07/—.	[FG returns to Leamington to attend STG who is fading.]	Leamington.
1844/08/—.	[FG attends STG.]	Leamington.
1844/09/09.	[FG and STG go to 57 Marina, St Leonards.]	St. Leonards-on-sea.
1844/09/18.	[STG supposes he has caught cholera.]	St. Leonards-on-sea.
1844/09/26.	[STG makes his last diary entry.]	St. Leonards-on-sea.
1844/10/23.	[STG dies at 1:30 PM.]	St. Leonards-on-sea.

1844/10/28.	[Corpse of STG is transported from St Leonards to Claverdon.]	St. Leonards-on-sea.
1844/10/31.	[Funeral of STG at Claverdon. The estate was worth £45,000 net.]	Claverdon.
1844/11/29.	[Last will and testament of STG is proved at London.]	London.

1845

1845/—/—.	[Galton is living at 105 Park Street in Mayfair, London, rooming with Frederick Waymouth Gibbs and Henry Robert Vaughan Johnson.]	
1845/01/08.	[FG and E. Galton attend a meeting of the Total Abstinence Society in Banwell, where FG 'strongly advocated the cause'.]	
1845/03/—.	[Galton is admitted to the Scientific Lodge No. 105 of the Freemasons at Cambridge, in the Third Degree.]	Cambridge.
1845/05/13.	[Adèle Galton marries Robert Bunbury, with FG present, at St. Mary's.]	Leamington.
1845/06/—.	[Elizabeth Galton is engaged to Edward Wheler.]	
1845/06/10.	[STG's 'superior furniture'[3] and 'cellar of rare and costly wines'[4] are sold by auction at 29 Lansdowne Place.[5]]	Leamington
1845/07/30.	[STG's extensive properties in the city of Birmingham are sold by auction.]	Birmingham.

3 Including bedsteads, a pair of Rosewood Loo Tables, a Spanish Mahogany Circular Dining Table, 15ft. 6in by 5ft with 5 loose leaves, 'on massive block, supported by 4 carved dolphins'.

4 Including Moselle, champagne, port, sherry, madeira, claret, brandy and Cambridge Ale (presumably the ale sent to him by FG).

5 *Leamington Spa Courier*, Saturday 07 June 1845, 2.

1845/09/28.	[Galton receives Greek visa valid till October 10.]	London?
1845/10/03.	[Letter from HF Hallam declining offer to join Galton in Africa.]	London?
1845/10/—.	[Diligence from Paris with Denham Cookes.]	Paris.
1845/10/—.	[Avignon, parts ways with Denham Cookes.]	Avignon.
1845/10/08.	[Galton receives visa at Marseilles.]	Marseilles.
1845/11/10.	[Sketch at Genoa.]	Genoa.
1846/—/—.	[Sketches of Verona and Albano.[6]]	Italy.
1845/11/—.	[Malta, meets Robert Temple Frere.]	Malta.
1845/—/—.	[Alexandria. Meets Montagu Boulton and Hedworth Barclay.]	Alexandria.
1845/11/25.	[Cairo at Shepheard's English Hotel. Firman received. Agree to go up Nile as a party.]	Cairo.
1845/12/31.	[Elizabeth Galton is married to Edward Wheler.]	

1846

1846/01/—.	[Galton, Boulton and Barclay are at Korosko by mid-January.]	Korosko.
1846/02/16.	[Galton celebrates his birthday at Khartoum with Mansfield Parkyns.]	Khartoum.
1846/04/—.	[Temple of Kom Ombo on the Nile. Philae (first cataract). Beni Hassan. Ibis.]	Nile.
1846/05/12.	[Galton receives visa, from the British Consulate at Alexandria, for Beirut.]	Alexandria.

6 They are annotated '1846' but there is no room in Galton's schedule for that; these are likely from late 1845, en route to Malta.

1846/05/13.	[Galton receives visa, from the French Consulate at Alexandria, for Beirut.]	Alexandria.
1846/06/05.	[At Beirut, sees British Consul Hugh Henry Rose.]	
1846/06/15.	[Sketch of hut at Damascus.]	
1846/06/21.	[At Damascus.]	Damascus.
1846/06/24.	[Letter from HF Hallam.]	Damascus.
1846/09/09.	[Letter from Montagu Boulton.]	Damascus.
1846/—/—.	[Letter from Montagu Boulton.]	Salayiyeh.
1846/09/30.	[Letter from Montagu Boulton.]	Beirut.
1846/11/16.	[Visa to travel from Damascus to Marseilles.]	Damascus.
1846/12/10.	[At Marseilles.]	
1846/12/—.	[Letter from HF Hallam.]	London.

1847

[Louisa's Summary] Lodgings in Pall Mall. Shot with [Fazakerley] in [Culrain], went on to Orkney.[7] Sir Hyde [Parker] hunted regularly [at] Stratton Lodge.

1847/—/—.	[Sketch of gateway to the Bishop's Palace at Liege.]	Liege.
1847/01/06.	[FG is present at the Warwickshire Hunt Ball in Stratford-on-Avon.]	Stratford-on-Avon.
1847/01/26.	[FG is at Mrs Torres' Annual Grand Dress Ball with 200 others.]	Leamington.
1847/02/27.	[FG is at Jullien's Concerts and Ball.]	Leamington.
1847/04/20.	[Letter from HF Hallam.]	
1847/04/23.	[FG attends the 'Shakespere Club Dinner'.]	Stratford-on-Avon.
1847/07/05.	[FG receives MA from Cambridge.]	

7 This is incorrect, as shown by Galton's letter to his mother from Culrain in 1848, making it clear that he was seeing it for the first time.

1847/08/22.	[Letter from Richard Partridge.]	London
1847/11/06.	[FG is at the Regent Hotel.]	Leamington.
1847/11/13.	[FG is at the Regent Hotel.]	Leamington.
1847/12/15.	[FG is a steward at the Infirmary Ball]	Leamington.

1848

[Louisa's Summary] Again Scotland. Shetland. Lucy died at Brighton, again Leamington. At Adèle's house & Lansdowne Hotel.[8]

1848/01/11.	[FG is at the Warwickshire Hunt Ball.]	Stratford-on-Avon.
1848/01/20.	[FG is at the Warneford Hospital Ball.]	Leamington.
1848/01/28.	[FG is at the Fifth Subscription Ball.]	Leamington.
1848/02/02.	[FG is at the Steeple Chase Ball.]	Leamington.
1848/02/05.	[FG rides with his brother Darwin and the Warwickshire Hunt.]	
1848/02/18.	[Emma visits Frank at his lodgings.]	London.
1848/04/10.	[Chartist convention at Kennington Common.]	
1848/05/31.	[FG receives visa for France.]	
1848/06/09.	[In Paris at the British Embassy.]	Paris.
1848/06/17.	[FG is at a meeting in Dee's Royal Hotel advocating a testimonial for Dr. Joseph Hodgson.]	Birmingham.
1848/06/23.	[Paris June Days uprising.]	
1848/08/28.	[Letter to Violetta Galton.]	Scotland.
1848/09/12.	[Letter from HF Hallam about grouse shooting in the Highlands.]	Culrain, Scotland.
1848/—/—.	[FG is in the Orkneys at Kirkwall, letter to Bessy].	Kirkwall.

[8] Presumably the one in Leamington.

1848/11/05.	[FG is present at the deathbed of his sister Lucy in Brighton.]	Brighton.
1848/11/11.	[Death of Robert Waring Darwin, father of the naturalist.]	

1849

[Louisa's Summary] Went to Shetland in early summer, returned with birds.

1849/03/08.	[FG is at the Bachelor's Ball]	Leamington.
1849/04/—.	[Report on Galton's character by the phrenologist Donovan.]	London.
1849/04/—.	[Letter re Telotype to Cambridge don.]	
1849/05/08.	[Visa for Norway.]	London?
1849/08/15.	[Fazakerley, Colonel Wallington and Paddy Johnson are shooting at Culrain]	

1850

South Africa.

1851

South Africa.

1852

1852/04/05.	[Galton returns to England from South Western Africa via St. Helena.]	Plymouth.
1852/04/26.	[At the RGS, Galton reads the second part of his paper on his explorations in Southern Africa.[9]]	London.

[9] Galton 1852.

1852/05/24.	[RGS Anniversary Meeting, addressed by Murchison, covering Galton's journey *inter alia*. Galton was almost certainly present.]	London
1852/06/14.	[Letter from William Jackson Hooker about care of plants.]	
1852/07/17.	[Hyde Parker on the schooner *Mayfly* (95 tons, launched Nov. 1851) sails from Portsmouth for Norway via Lowestoft, Norfolk, accompanied by George William Pierrepont Bentinck (1803–1886) in the yawl *Dream*. Bentinck, known as 'Big Ben', was standing for the Tories as an MP for the Norfolk-West constituency in the general election of July 1852; he was elected.[10]]	Portsmouth.
1852/07/31.	[The *Mayfly* and *Dream*, along with the Liberal M.P. Thomas Milner Gibson (1806–1884)[11] in the 65 ton schooner *Sapphire*, sail from Lowestoft, Norfolk for Norway.]	Norfolk.

[10] 'Mr. Bentinck was the very type and model of the old school of yachtsmen who eschewed racing, loved the sea for its own sake, lived chiefly on their vessels, and carried the tradition of the first commodore into recent times. ... He had the greatest contempt for people who did not live on board their vessels, employed captains or sailing masters, and who confined their yachting to the safe waters of the Solent. ... He was always delighted to take friends on a sea voyage, but could never be induced to give any particulars as to where bound or the probable length of the cruise, and very much resented an inquiry on either point. People accordingly who accompanied him always settled their affairs for a reasonable period, not knowing when they would return. Long voyages and the difficulties of bad weather were Mr. Bentinck's delight ... As a confirmed bachelor he resented the prominence given to feminine society in the later days, and was accustomed to speak of the leading spirits of the new order as "captains of foot." Mr. Bentinck was well known in wider circles, and his opposition to the prevailing spirit of progress was the delight of session after session in the House of Commons.' Guest 1903, 293–295.

[11] Gibson was 'one of the relatively few yachtsmen who could sail and navigate his own vessel' and 'the last yachtsman to sail in the Mediterranean under a pass from the Dey of Algiers', Guest 1903, 69.

1852/08/15.	[Gibson arrives in Kirkwall Bay, in advance of Hyde Parker and Bentinck, having parted from them in the Orkneys.]	Scotland.
1852/09/06.	[Galton's paper 'Expedition to the Eastward of Walfisch Bay' is read at the British Association in Belfast by a Professor M'Douall.]	
1852/09/23.	[Hyde Parker on the *Mayfly* returns from Norway and the west of Scotland. G. Bentinck returns in the *Dream*.]	Scotland.
1852/11/28.	[Galton attends the funeral of the Duke of Wellington at St Paul's, along with over a million Londoners, but is overcome by illness.]	London.
1852/12/—	[Galton is at Claverdon with his mother, still ill.]	Claverdon.
1852/12/30.	Thu. Came to Dover [32 Marine Parade[12]].	Dover.[13]
1852/12/31.	Fri. Dined at Winthrops. Minet's Ball.[14]	Dover.

1853

[Louisa's Summary] Left Dover in March. Engaged in April. Spirit rapping mania. Married in August.

1853/01/01.	Sat. Mrs Wellden's Party.	Dover.
1853/01/03.	Mon. Jekens eveng. party[15]	
1853/01/04.	Tue. Ball on the heights.	
1853/01/06.	Thu. Butler's eveng. party.	
1853/01/07.	Fri. Maddock's ball.	

12 Emma's Diary.

13 Galton, Harwood's Diamond Diary 1853.

14 The Minets were Huguenots, long-established in Dover. Benjamin Eveleigh Winthrop (?–1870) was a barrister on Marine Parade.

15 The Jekens family were brewers.

1853/01/08.	Sat. Campbell came to Dover. Dined at Ship Hotel.[16]
1853/01/09.	Sun. Walked to Deal.[17] Campbell left.
1853/01/10.	Mon. Butler's Even. Party.
1853/01/13.	Thu. Mrs [Girandot] Evening party.
1853/01/14.	Fri. Butlers Evening Party.
1853/01/17.	Mon. Russell's concert.
1853/01/18.	Tue. Mrs Minet Eveg. Party
1853/01/19.	Wed. Move into new house 6 Waterloo Cresc.[18] Taken house for 6 weeks.
1853/01/20.	Thu. Mr Eveleigh Winthrop's Ball at [Warner's Rooms] Cliff Fell
1853/01/21.	Fri. Dine Steward, on Heights, 7h.
1853/01/25.	Tue. Mrs Hay Evg. Party. We give a dinner party. Capn. Gibbon slept here.[19]
1853/01/27.	Thu. Practise the Parisienne.
1853/01/28.	Fri. Dine Mrs Minet. Public Ball. Jekens' dinner refused.
1853/01/31.	Mon. Mrs. Butler & Mrs. Ackers
1853/02/01.	Tue. Mrs Jekens Eve.
1853/02/03.	Thu. Mrs. Jekens dine $\frac{1}{4}$ to 7.
1853/02/04.	Fri. Miss Ibbotson Eve P. 8h30. Dined at the Maddock's. Refused dining at the Castle.
1853/02/07.	Mon. Mrs. Latham Eveg.
1853/02/08.	Tue. Butlers E. Py.
1853/02/10.	Thu. Mrs Rennell dine $\frac{1}{4}$ to 7.

[16] On Custom House Quay.

[17] Eight miles from Dover.

[18] Just off the Marine Parade.

[19] Evidently not the historian Edward Gibbon, who served with the Hampshire Militia as a Captain and was at Dover in 1762.

1853/02/12.	Sat. A small dinner at home of Mr Winthrop. Mrs Minet Captn. Parker Miss [Butler[20]] Sent book (1st part) off to Murray.[21]
1853/02/12.	Frank sent off part of his book, book finished March 1st.[22]
1853/02/14.	Mon. Dine Miller on the heights 7. Butlers Eveg. P. Refused Larking's dinner.[23]
1853/02/15.	Tue. Mrs. Wellden's dinner $\frac{1}{4}$ to 7. Refused Mr. Winthrop's dinner.
1853/02/16.	Wed. Mrs. Smith's dance.
1853/02/17.	Thu. Our dance. Dine at the Minets. Capt. Gibbon comes.
1853/02/18.	Fri. Mrs Girandot's Eve P.
1853/02/21.	Mon. Baroness Eve P. Dined at the Minets.
1853/02/22.	Tue. Minets theatricals.
1853/02/24.	Thu. Refused Moloney's dinner. Minets theatricals.
1853/02/25.	Fri. Webbes 7 [o'clock] dine
1853/02/28.	Mon. Butler's Eve P. Wrote stories.
1853/03/01.	Tue. Baroness Eve Party.
1853/03/03.	Thu. Baroness dine $\frac{1}{4}$ to 7.
1853/03/04.	Fri. Walk to Kearney Abbey.
1853/03/05.	Sat. Larking's dinner $\frac{1}{4}$ to 7.
1853/03/08.	Tue. Came back to London.
1853/03/09.	Wed. First part of book came to be corrected.

[20] Possibly Baker.

[21] Galton 1853.

[22] Emma's Diary.

[23] John Wingfield Larking (1801-1891), a merchant at Alexandria. Burton appears to have met Larking at Dover too, though he later claimed, perhaps for dramatic effect, that he only met him on the boat to Egypt some months after this, en route to Mecca.

1853/03/10.	Thu. Maddock Eve Party. Russell's dinner $\frac{1}{4}$ to 7. Baroness Eve Party.
1853/03/11.	Fri. Du Vernets Dinner. Walk beyond Shakespeare cliff.
1853/03/15.	Tue. Butler's Eve Party
1853/03/16.	Wed. A. E. Broadley married to Mr. Sykes.[24]
1853/03/17.	Thu. Dine at the Baroness de Sternberg.[25]
1853/03/22.	Tue. Left.
1853/04/17.	Sun. Dine with Campbell at the Plough.
1853/04/17.	Heard Frank had gained the Gold Medal.[26]
1853/04/18.	Mon. Mrs Hollond Evening 9. [Kew] At $12\frac{1}{2}$.
1853/04/19.	Tue. Greenwich Ball, go first to Lady Dalyell. Dine Moira $7\frac{1}{4}$. Call Hollond 5 $\frac{1}{2}$.
1853/04/20	Wed. Mr. Edmonds will come at 11 to me.
1853/04/21.	Thu. Dine Mr. Hallam $7\frac{1}{2}$. Goldsmiths ball.
1853/04/23.	Sat. Ev with Mrs Hollond to Crystal Palace. Ld Rosse.[27] Heard from the Dean[28] ...
1853/04/24.	Sun. Called at Mrs Hollond's &c.&c.&c.
1853/04/25.	Mon. Mrs Hollond dine $\frac{1}{4}$ to 7. Called in morning &c.&c.
1853/04/26.	Tue. &c.&c. Dined with Campbell at Greenwich.
1853/04/28.	Thu. Mr. Larking to dine 7.

24 Anne Elizabeth Broadley married Joseph Sykes of Raywell, Yorkshire. She was a sister of Fanny Broadley, who was friendly with Galton.

25 Catherine Augusta Harrison (1791-1859).

26 Emma's Diary.

27 William Parsons (1800-1867), third Earl of Rosse in Ireland.

28 George Butler (1774-1853), Dean of Peterborough and the father of Louisa Butler.

1853/04/29.	Fri. Started very early for Hatton & Claverdon, dined at Edstone, my mother sending me in the Brougham.
1853/05/01.	Sun. Went to eve service in the cathedral. Walked out a little.
1853/05/02.	Mon. Walked with G. A. S. Butler to Milton Park.
1853/05/03.	Tue. Returned from Peterbro'.
1853/05/05.	Thu. Dalyell & I dined together at the Blue Posts after a long walk through Kensington Gardens with Stander
1853/05/06.	Fri. Dalyell Holway & I dined at Billingsgate in evening. Met Clarke Simpson, Spring Rice & several others in smoking room—quite a gathering of old friends. Darwin[29] came to town.
1853/05/07.	Sat. Ld Rosse Soiree.
1853/05/08.	Sun. Wet evg. went with Campbell to the Plough.
1853/05/09.	Mon. Clarke and I went to hear the Duke of Northumberland examined & then to the spirit rappings wonderful. Mrs Hollond, Lady Radstock. Geograph. Lady Stratheden[30] return
1853/05/10.	Tue. Mrs Barlow $7\frac{1}{4}$ to dine. Hollond's Evg.
1853/05/14.	Sat. Boulton dine refused. [Left] to go to Gayton.
1853/05/17.	Tue. Left Gayton at 7 a.m. Lady Radstock.
1853/05/23.	Mon. Geograph. Anniversary & Dinner. Prideaux — Mrs Hollond.

[29] His brother Darwin Galton

[30] Presumably the mother of his Cambridge friend F. Campbell.

1853/05/24.	Tue. Dine Boulton $6\frac{1}{2}$. Opera. Spirit rapping 4th time.
1853/05/25.	Wed. Derby Day.
1853/05/26.	Thu. Mrs Hutton Putney Dance.[31]
1853/05/27.	Fri. Murray (publisher) dine refused.
1853/05/28.	Sat. Ld Rosse
1853/05/31.	Tue. Dine Boulton $7\frac{1}{2}$ or $\frac{3}{4}$.
1853/06/01.	Wed. Breakfast Monckton Milnes 10h. 16 Upper Brook St. French play Campbell. [dinner] Boulton refused.
1853/06/02.	Thu. Dine Mrs Phillips.
1853/06/03.	Fri. Admiral Hyde Parker dine $\frac{1}{2}$ past 7 Admiralty. Hollond dine (refuse).
1853/06/06.	Mon. Mrs. Burchell Evg. Party. 42 Upper Harley St.[32] Mrs Hollond.
1853/06/07.	Tue. Go to Wembley by the 11h NW Train to Sudbury. Mr Milnes evening. Dine Townall $6\frac{1}{2}$ 63 Guilford Russell Sq.
1853/06/08.	Wed. $\frac{1}{4}$ past 4 at Barclays bank to coach with Gurney Barclay to Mrs. Hayden.[33] $9\frac{1}{4}$ breakfast Farnaby Cator.
1853/06/09.	Thu. 12 Mrs. Hayden.
1853/06/10.	Fri. 12 Mrs. Hayden. 2. Boulton.
1853/06/20.	Mon. Geograph.
1853/06/21.	Tue. Chobham camp.[34] Dine Douglas $7\frac{1}{2}$.

[31] Crompton Hutton, Putney Park.

[32] Mrs. William Burchell.

[33] Mrs. Maria B. Hayden, the medium and spirit rapper.

[34] Large-scale military preparations in Surrey, attended by many spectators.

1853/06/22.	Wed. Breakfast M. Milnes $10\frac{1}{4}$ Dine David Barclay $7\frac{1}{2}$ Regents Park.
1853/06/23.	Thu. Lady Radstock Eve. Dine Aldersley at Coventry at 8. Breakfast Douglas $8\frac{3}{4}$.
1853/06/24.	Fri. Dine Prideaux $7\frac{1}{4}$ Goldsmith Hall. Breakfast Dr. Hodgkin 9.[35]
1853/06/28.	Tue. Dine Sir C. Cunningham. 4 Clayes St. 8 o'clock.
1853/06/30.	Thu. Dine Boulton [Sand...]
1853/07/01.	Fri. Dine Howard. Windh. 7.
1853/07/02.	Sat. Dine Murray refused. Peterborough go to.
1853/07/07.	Thu. Mrs Batten 7 dine. 21 Westbourne Terr.
1853/07/08.	Tea-ed Gurney
1853/07/09.	Dine Douglas $7\frac{1}{2}$.
1853/07/13.	Lady Hollond Evg.
1853/07/25.	[Letter from Charles Darwin re *Tropical South Africa.*]
1853/08/01.	Frank married Louisa Jane Butler.[36]
1853/08/—	[Frank and Louisa leave for Switzerland and Italy, returning only in March of the following year.]

[Note on last page] 22. Queen Anne Ct.
Mr Thomas Allen
Mrs Heywoods Museum
Batts
Coventry
John Allen's brother

Cooper wood cats. Dalyell. The friends of the Dean of Peterborough. So many old pupils.

35 Possibly Dr. Thomas Hodgkin (1798–1866).
36 Emma's Diary. They were married at the cathedral in Peterborough.

Captain Younghusband book. (a) North and his daus.

1854

[Louisa's Summary] Returned to England in March by Mt Cenis to Paris. Went to Leamington. Lodgings in Portugal street. Came to Victoria street in July. Went abroad Aug 6th Rouen. Chambord. Blois. Nantes. Dinan. Avranches. Fever. Mr Andersson joined us.[37]

1854/03/—	[Frank and Louisa return from Italy, having wintered in Rome and Florence.]	England
n.d.	Campbell at Cambridge	
1854/05/01.	Mon. Mrs. Hollond $7\frac{1}{4}$.	
1854/05/02.	Tue. Mrs Aytoun 7.	
1854/05/03.	Wed. Levee. Lushington[38] early.	
1854/—/—.	[Letter from Mansfield Parkyns.]	
1854/05/06.	Sat. Dalyells 7h.	
1854/05/11.	Thu. Lushingtons $7\frac{1}{2}$ h.	
1854/05/12.	Fri. Ingleby[39] at 11. Wembley.	
1854/05/13.	Sat. Launch R. Albert.[40] Oxford.	
1854/05/15.	Mon. Emma comes.	
1854/05/16.	Tue. Dine [P Erle] $7\frac{1}{2}$.	
1854/05/19.	Fri. Dine Batten 7. Darwin refuse. Sir G. Back refuse.	
1854/05/20.	Sat. Cambridge.	Cambridge.
1854/05/22.	Mon. Geogr. Anniversary at one. 21 Regent Sq. Geogr. Dinner 7. Free Mason's tavern.	London.

37 Galton's entries are from the Engagement Book for 1854. No clear months/dates. Consecutive entries with day of month numbers. Dated 1854 by entry for the launch of the *Royal Albert*.

38 Almost certainly Franklin Lushington (1823-1901) who had been at Trinity College with Galton.

39 Colonel Ingilby, who had travelled on the *Dalhousie* to the Cape with Galton and Andersson.

40 The screw-propulsion ship *HMS Royal Albert* was launched on May 13, 1854 at Woolwich.

1854/05/23.	Tue. Dennisons 7.	
1854/05/24.	Wed. Mrs Butler dine at $6\frac{3}{4}$.	
1854/05/26.	Fri. Eras. Darwin 7	
1854/05/27.	Sat. Cologne Band.	
1854/05/28.	[Letter from Charles Darwin on fire-making.]	
1854/05/31.	Wed. Emma comes for a week or ten days.	
1854/05/31.	To Frank and Louisa in Portugal Street; moved in July to 55 Victoria Street.[41]	London.
1854/06/01.	Thu. Mrs Charles Butler dine $7\frac{1}{2}$. H Lefevres.	
1854/06/02.	Fri. Brougham 2 or $2\frac{1}{2}$.	
1854/06/03.	Sat. Brougham from 20 to 4 to $\frac{1}{4}$ past 6. Mrs. Aytoun $\frac{1}{4}$ 8.	
1854/06/05.	Mon. Mr Hollond.	
1854/06/06.	Tue. Darwin.	
1854/06/07.	Wed. Darwin.	
1854/06/09.	Fri. Levée. Carriage came $1\frac{1}{4}$. Sent away $5\frac{3}{4}$.[42] Mrs Kay.	London.
1854/06/10.	Sat. Crystal Palace.	London.
1854/06/12.	Mon. Geogr. Council. Lunch at 1. Leave [H...] Geograph.	London.
1854/06/13.	Tue. [Evening] M. Milnes.	
1854/06/14.	Wed. Ld Ellesmere.	
1854/06/15-17	Thu-Sat. Darwin.[43]	
1854/06/14	Mon. Dine Lefevres. Green St.	
1854/06/20.	Tue. Dinner at home 7.	London.
1854/06/21.	Wed. Mr Rasch $7\frac{1}{4}$. Ld Ellesmere.	
1854/06/22.	Thu. Mr Hallam. 7.	
1854/06/24.	Sat. Harrison comes.	

[41] Emma's Diary.

[42] Galton was presented to Queen Victoria at her afternoon levee, by Lord Ellesmere.

[43] Presumably his brother Darwin Galton.

1854/06/26.	Mon. Harrison goes. Mrs Joyce comes. Hollonds $7\frac{1}{4}$.	
1854/06/27.	Tue. John Lefevre's $9\frac{1}{2}$.	
1854/06/28.	Wed. Mrs Joyce goes.	
1854/06/29.	Thu. Harrow speeches. Mrs Aytoun.	
1854/06/30.	Fri. Crystal Palace.	
1854/07/01.	Sat. Stanmore.	
1854/07/03.	Mon. Cab to Campbell at Brook at $4\frac{1}{2}$. Greenwich. Geographical.	
1854/07/04.	Tue. Keep open for Mr Arch: Smith[44] R. Gurney $7\frac{1}{4}$.	
1854/07/06.	Thu. Mr Westby. $7\frac{1}{2}$.	
1854/07/19.	Wed. Council Geograph. 3 PM.	London.
1854/07/21.	Fri. Dine Boulton at 7.	
1854/08/06.	[Galtons leave for France]	
1854/09/27.	[C.J. Andersson joins the Galtons at Dinan in France]	France
1854/10/—.	[Galtons have returned to London by way of Boulogne in late October, leaving Andersson in Avranche]	London
1854/11/27.	[Galton offers remarks at the RGS on Livingstone.]	London.
1854/12/—	[*Art of Travel* published by John Murray.]	

... If the number of I's is the test of degree
Lord Brougham I acknowledge your station
Your works egotistical warrant your plea
For what they imply they themselves guarantee
That you're the first star of the nation.[45]

1855

[Louisa's Summary] Leamington in March. Edstone & Hadzor. Frank went to Aldershot in July. Oriel Cottage. We

[44] Likely Archibald Smith (1813-1872) the mathematician and lawyer, who had been at Trinity College, and Senior Wrangler in 1836.

[45] Enclosed scrap of paper.

went to Paris in September with Emma & Gertrude. Xmas at Leamington. Frank lectured at the camp till spring.

[Galton's Notes] Fontainebleau in 1855 (after Camp). Lord Ashburton Jan. Barford bridge. Umbrella tent. Bristed. Lomardy[46] (I was at H. Castiglione[47] Leon my servant). I think Gertrude was with us. We were 43 days abroad. Mr [A...]. Mr Hahn lunch. Rev H Hahn of Barmen. Letter from John Allen & Hans Larsen Dec 14 / 1854. Next address at Melbourne. Medal to Charles J. Andersson.[48] Mrs Somerville.[49] [...]. Dinner at Hollond Gurney Rae Percival Spencers.

1855/01/01.	[Letter from Charles Darwin about *Art of Travel.*]	London.
1855/01/08.	[Galton is present at the first meeting of the RGS in their new premises at Whitehall Place. A paper on the safety of Dr. Barth was read.]	London.
1855/01/09.	[Emma] Frank's "Art of Travel."[50]	London.
1855/02/12.	[Galton is present at RGS general meeting.]	London
1855/02/22.	[Letter from Charles Darwin re domestication.]	
1855/02/26.	[Galton is present at RGS general meeting.]	London
1855/03/—.	Leamington in March.	Leamington.
1855/04/04.	[Letter from Charles Andersson.]	
1855/04/06.	[Letter from Charles Andersson.]	
1855/04/16.	[Letter from Charles Andersson.]	
1855/05/03.	[Emma] To Frank and Louisa to 14th.[51]	London.

46 Lombardy
47 A village in Lombardy.
48 Andersson actually received surveying instruments rather than a medal.
49 Mary Somerville (1780-1872), who wrote a book on *Physical Geography* (1848).
50 Emma's Diary.
51 Emma's Diary.

1855/05/23.	[Galton is present with Carl Hugo Hahn at the 'Geographical soirée' held at Lord Ellesmere's Picture Gallery.]	
1855/05/28.	[French Geographical Society awards Galton their Silver Medal for his South African expedition. Galton arranges that Andersson is awarded instruments by the RGS.]	London.
1855/06/23.	[Emma] Frank's appointment to Aldershot.[52]	London.
1855/07/—.	Frank went to Aldershot in July.	Aldershot.
1855/07/02.	[Galton is appointed 'Instructor in Common Things' to the Militia regiments camped at Aldershot. Lives nearby at Oriel Cottage.]	
1855/07/10.	[Letter from Charles Andersson.]	
1855/07/11.	[Letter from Charles Andersson.]	
1855/07/13.	[Letter from Charles Andersson.]	
1855/07/19.	[First Lecture at Aldershot.]	
1855/07/28.	[Second lecture at Aldershot, Lighting Fires]	
1855/07/30.	[Repeat of second lecture at Aldershot.]	
1855/08/01.	[Third lecture at Aldershot: Water.]	
1855/08/02.	[Repeat of third lecture at Aldershot.]	
1855/08/04.	[Fourth lecture at Aldershot: Bivouacs.]	
1855/08/06.	[Repeat of fourth lecture at Aldershot.]	
1855/08/09.	[Galton's pamphlet *Arts of Campaigning* is published by John Murray.]	

---

52 Emma's Diary.

1855/08/11.	[Fifth lecture at Aldershot: Food.]
1855/08/13.	[Repeat of fifth lecture at Aldershot.]
1855/08/15.	[Sixth lecture at Aldershot: difficulties on the march.]
1855/08/16.	[Repeat of sixth lecture at Aldershot.]
1855/08/18.	[Seventh lecture at Aldershot: Crossing water and broken ground.]
1855/08/20.	[Repeat of seventh lecture at Aldershot.]
1855/08/22.	[Eight lecture at Aldershot: Carpentry.]
1855/08/23.	[Repeat of eight lecture at Aldershot.]
1855/08/25.	[Ninth lecture at Aldershot: dressing hides.]
1855/08/27.	[Repeat of ninth lecture at Aldershot.]
1855/08/29.	[Tenth lecture at Aldershot: soap etc.]
1855/08/30.	[Repeat of tenth lecture at Aldershot.]
1855/09/01.	[Eleventh lecture at Aldershot: writing and drawing.]
1855/09/03.	[Repeat of eleventh lecture at Aldershot.]
1855/09/03.	[Emma] To F. and L. at Farnborough.[53]
1855/09/05.	[Twelfth lecture at Aldershot: matting and cloth.]
1855/09/06.	[Repeat of twelfth lecture at Aldershot.]
1855/09/08.	[Thirteenth lecture at Aldershot: draught animals.]

[53] Emma's Diary.

1855/09/10.	[Repeat of thirteenth lecture at Aldershot.]	
1855/09/12.	[Fourteenth lecture at Aldershot: hutting and tenting.]	
1855/09/13.	[Repeat of fourteenth lecture at Aldershot.]	
1855/09/—	We went to Paris [Exhibition, at the Palace of Industry] in September with Emma & Gertrude. Fontainebleau. Castiglione [in Lombardy].	France
1855/10/—	[Galtons return to London from France in mid-October]	.
1855/11/09.	[*Cambridge Essays*, containing Galton's 'Notes on Modern Geography', is published.]	
1855/11/11.	[Galton is present at RGS general meeting.]	London.
1855/12/10.	[Galton is present at RGS general meeting.]	London.
1855/12/—	[Second (revised and enlarged) edition of *The Art of Travel* is published, but dated 1856.]	
1855/12/—	Xmas at Leamington.	

1856

[Louisa's Summary] To the Norths at Hastings at Easter. Fever & chin bad in April. Ella's concerts. Ailing all summer. Went abroad in June. Emily with us. Tour in Switzerland, the Tyrol & North Germany. Met Monty at Dresden, returned in Sep. October & went to Leamington. Teeth out in December. Josephine delicate. Ordered to leave Oxford by Sir J. Clarke. They fix upon Clifton. Catharine goes with them to settle them.

[Frank's Summary] 1856. Art of Travel
(a) Innsbruck July 27 with Emily[54] [...] Bolzen Coln Bonn Heidelberg [T...] S. Moritz

[54] Possibly Emily Butler.

Mrs Hone's address (Mr Hahn's sister)
c. Barth's paper
Palliser[55]
Lodgings Mortlake from April 17
Lessops
Crawfurd
d. tour in Tyrol with Emily 111 days started June 28

1856/01/01.	Lunch at Rylands.
1856/01/16.	FG first lecture.[56]
1856/01/20.	Met Dr Barth. Erhardt[57] stopped.
1856/02/04.	[Letter from Charles Darwin re Erhardt and domesticated poultry.]
1856/02/07.	Bentham dine.
1856/02/09.	FG to Cambridge.
1856/02/11.	Called on Mrs [Barlow].
1856/02/16.	Dined Sir J Stephen[58] met [V. Johnson[59]] & Gibbs. Called Mrs du Plat.
1856/02/20.	F. went to the camp.[60]
1856/02/21.	Lefevres[61] to lunch and to Munro's studio.[62]
1856/02/22.	Early lunch with [Eales].
1856/02/29.	Lord Dungannon to lunch.[63]
1856/03/09.	FG had bronchitis for 5 days.
1856/03/28.	To Norths at Hastings.
1856/04/01.	[Took] lodgings.

55 The Irish explorer Captain John Palliser (1817–1887), who journeyed in North America.
56 At Aldershot, for that year.
57 Johan Jakob Erhardt (1823–1901), a missionary for the Church Missionary Society, who had been based in East Africa with Johannes Rebmann and brought back reports of a fabulous inland lake there.
58 James Fitzjames Stephen.
59 H. Vaughan Johnson, who had been at Trinity with Galton.
60 Aldershot.
61 Sir John George Shaw-Lefevre (1797–1879).
62 Possibly R. Munro, the instrument manufacturer.
63 Arthur Hill-Trevor (1798–1862).

1856/04/05.	[Letter to *The Times*, 'Signals Available to Men who are Adrift on Wrecks at Sea.'[64]]
1856/04/08.	Left for London
1856/04/11.	FG lecture[65]
1856/04/14.	Drive with Mrs Sabine
1856/04/18.	FG's last lecture.[66] LG ill. jaw. fever.
1856/04/23.	FG to Naval Review.
1856/05/02.	LG chin still bad.
1856/05/03.	FG to [work] at [Cadfield].
1856/05/16.	Clives.
1856/05/17.	To Waverley.
1856/05/20.	Mrs Hollond.
1856/05/24.	['Jamaica: Interesting Discovery' in *Notes and Queries*.[67]]
1856/05/26.	London, Victoria Street till June 13th.[68]
1856/05/29.	Fireworks for the peace.[69] 3 sets of Galtons to see them from our house.
1856/06/—.	[Calling card from William Sandys Wright Vaux]
1856/06/04.	Party at home. [Hallams] Stephens Clives &c.
1856/06/05.	Dined R Gurneys.

64 Galton 1856b.

65 To the United Service Institution, on 'The Arts of Travelling and Campaigning, Hutting and Tenting'.

66 His second and last lecture to the United Service Institution. 'He showed the best mode of tent-making and tent-planting, preserving meat, making fire, seasoning green wood, &c. &c., under difficulties. Mr. Galton hinted that a traveller might even make his own paper, as well as his soap and candles, but time did not allow him to enter into details on these points. The lecture was well attended by an interested audience, including Lord Radstock, General Bagnold, Hon. Col. Bruce, General Ferguson, Colonel Matheson, Colonel Alcock, Hon. Captain Murray, R.N., &c.' *Worcestershire Chronicle*, Wednesday 23 April 1856, 2.

67 Galton 1856a.

68 Emma's Diary.

69 Peace with Russia. £10,000 was spent by the government on fireworks.

1856/06/06.	At Clives.
1856/06/07.	Geological lessons.
1856/06/14.	To Wembley.
1856/06/28.	Started for abroad from St. Cath. wharf for Ostend Bruges Liege Cöln Bonn to Drackenfels with Adèle & [...].
1856/07/05.	[B...] Heidelberg. chin painful. Lucerne Horgen Thusis St. Moritz Samedan Nauders Landeck.
1856/07/27.	Innsbruck FG busy with glitter. Berchtesgaden Salzburg [...].
1856/08/06.	[Hollonds] [...]
1856/09/12.	Linz.
1856/09/13.	Vienna.
1856/09/16.	FG to the Naturforscher.
1856/09/19.	Prag[ue].
1856/09/22.	Dresden. Monty breakfasted with us.
1856/10/04.	Saxon Switz:
1856/10/12.	Coln ([...])
1856/10/17.	Home.
1856/10/19.	Mr Brewin's funeral.[70] dined Nat Phillips - [Edstone & home] then Edstone again.
1856/11/11.	Day at Gayton & to London.
1856/11/14.	To Warley.
1856/11/17.	Saw Brodie.[71]
1856/12/09.	Tooth out by Cartwright.
1856/12/12.	Dined Archie Smiths. FG to hear Spurgeon.[72]
1856/12/18.	2 more teeth out.
1856/12/25.	Josephine was to leave Oxford.

[70] Charles Brewin, who had married Frank's Aunt Sophia, had died on Oct. 12.

[71] Possibly Sir Benjamin Brodie (1817–1880).

[72] Charles Haddon Spurgeon, the evangelical preacher.

I do not at all think that adequate attempts have ever been made to obtain a free African immigration. The number of recruits depend on the skill of the recruiting officer. We must ingratiate ourselves more with the African tribes generally.

1857

[Louisa's Summary] Austen Charles born in May. My teeth replaced towards the close of March. Frank as an indemnification for the dull winter took me to Paris at Easter, we were about a fortnight, I spent five days at Fontainebleau. Gave up the flat in the summer. Visit to Edstone early in July. Frank photographed in the cattle wagon, then spent a few days at Leamington & started for Switzerland in August. Went to Geneva, Chamounix, Cormayeur, where staid ten days with the Russell Gurneys. Orta & Bellagio, where met Arthur & Montague, shortly after, they went to Greece; then to Milan, Genoa & home by the Corniche. Heard of Amie's death on my return to Paris, took possession of Rutland Gate, furnished. Catherine engaged to be married Nov 10. The marriage took place Nov 24. They sailed Nov 26. Xtmas in Devonshire Terrace. George & Josephine there.

1857/—/—.	[Fever prescription for Galton from Dr. David Livingstone. A concoction of quinine, calomel [mercury chloride], jalap resin and cardamom tincture.]
1857/01/—.	Surrey Chapel to hear Spurgen[73] Bentham.
1857/01/17.	Took possession of our lodgings (?Norwood).
1857/01/24.	Returned.
[n.d.?]	Vaughan John.[74] Mr Aytoun. Norths to tea. Capt Palmer to dinner. [Joyces].
1857/01/31.	Mrs Aytoun - Fanny Denison.
1857/02/18.	FG dined at Sabines.
1857/02/19.	With Mrs Aytoun.

[73] Charles Haddon Spurgeon.

[74] Vaughan Johnson.

1857/03/16. Spottiswoodes in evening.

1857/03/28. Dined D. Barclays.

1857/04/01. To Fontainebleau to 9th. Home about 13 (dined regularly at the Geographical).

1857/04/01. Left for Paris & Fontainebleau 4th to 9th.

1857/04/10. [Letter from W. G. Clark.]

1857/04/15. April back home.

1857/04/24. FG to Palliser.

1857/04/29. Dinner party Stephens, Maines, [Sinjans].

1857/05/08. Bentham to dinner.

1857/05/12. [Saw] Russell's ship at Blackwall (Gt. Eastern).[75]

1857/05/12. [Letter from Heinrich Barth.]

1857/05/16. With Bentham blue bells at Kew.

1857/05/17. [Letter from Charles Babbage about *Art of Travel*.]

1857/05/21. London to June 12th, Frank and Louisa, Victoria Street.[76]

1857/05/24. Mr Latrobe with us.

1857/06/03. FG up the river with Douglas.

1857/06/05. [Letter from Rev. Joseph Williams Blakesley.]

1857/06/09. Barth to Ascott.

1857/06/09. Dr Barth to breakfast.

1857/06/13. Fixed to take Rutland Gate.

1857/06/13. L. fixed to take Rutland Gate.

1857/06/14. W.G. Clark & Spottiswoode to breakfast.

1857/06/14. Frank bought 42 Rutland Gate for £2,500, gas extra £35.[77]

[75] Scott Russell was building the Great Eastern ship.
[76] Emma's Diary.
[77] Emma's Diary.

1857/06/27.	[Letter from Stephen Lushington.]	
1857/07/04.	Ockham.	
1857/07/05.	To Ockham.	
1857/07/07.	[Letter from Charles Darwin about domestication of animals.]	
1857/07/11.	Edstone. FG photographed. Mrs. Burton & Miss Venn.	
1857/07/11.	To Edstone.	
1857/07/17.	The pond dragged.	
1857/07/28.	[Letter from Robert FitzRoy re Royal Society.]	
1857/07/30.	Dieppe. Paris. Chamonix. Courmayeur	
1857/07/30.	Left for Dieppe - Koln - Chamonix - Col de [Bonhomme].	
1857/08/02.	[Letter to Richard Burton. Saw Jomard in Paris.]	France
1857/08/13.	Courmayeur.	
1857/08/14.	[Galton and Leslie Stephen unsuccessfully attempt to climb Col du Géant.]	
1857/08/19.	FG & Stephen to Col du Géant [succeeding this time].	
1857/08/19.	FG - Stephen Col du Géant, Chatillon [G...] Varallo Orta Jenkinsons Arena Lugano.	
1857/09/04.	[Varallo]	
1857/09/05.	[Orta ...]	
1857/09/11.	Signor Cattaneo Bellagio - Artiz [Moritz].	
1857/09/14.	Lugano.	
1857/09/18.	Met both Arthur & Monty at Bellagio.	
1857/09/21.	[Letter from Charles Andersson.]	
1857/09/25.	Milan. Diligence to Nice. Arles.	
1857/10/—.	[Galton is elected to the Geographische Gesellschaft, Vienna]	

1857/10/16	Rutland Gate.
1857/11/08.	FG to Campbell of Islay.
1857/11/13.	FG breakfast with Dr. Barth.
1857/11/24.	Catherine[78] married Dr. Bowen.
1857/12/03.	To see Gt. Eastern at Greenw.[79]
1857/12/17.	Mr. Atkinson & Spottiswoode to dinner.
1857/12/24.	['The exploration of arid countries' in PRGS.[80]]
1857/12/26.	[Letter to the *Times*, 'Negroes and the Slave Trade'[81]].

[At some stage in 1857, Galton presented models, constructed to illustrate the 'Arts of Camp Life', to the military academy at Woolwich. These may have been the models used at Aldershot in 1855-6, or they may have been specially made.]

1858

[Louisa's Summary] George went to Cheltenham in January as Vice Principal of the college. Frank busy all the winter superintending the models for the Art of Camp Life.[82] At Easter we went to Hastings to stand sponsors for Amy's baby Sophy Bree[83] & on to Leamington. The Biggs & Maggie came to us in May, then Emma: drew with her at the Vernon Gallery, left London the middle of July. Spent an evening at Gayton, the next day to Grasmere, where passed a very pleasant month (the Norths there) then to Keswick where had 3 weeks of bad weather. Fatal news of dear Catherine by the African mail in Sept, I heard it on the 8th.[84] Mamma in Switzerland with the sisters & Montagu. British Association Leeds. I went to Maggie. Frank's Heliostat brought out. Met Mamma at Malvern

[78] Louisa's sister.

[79] The steamship SS Great Eastern, designed by Brunel, then under construction in the shipyards opposite Greenwich.

[80] Galton 1857b.

[81] Galton 1857a.

[82] Galton 1858.

[83] Galton's great-niece, a daughter of Lucy Amelia Moilliet.

[84] Her sister Catherine had married Rev. John Bowen, the Bishop of Sierra Leone. She died after childbirth on 1858/08/04 in Freetown. See Edward Graham *The Harrow life of Henry Montagu Butler* (Longmans, 1920).

in October, remained at Wick; a few days before Xmas for Edstone staid there a week, then to Bertie Terrace for a week. Darwin fell ill of Diphtheria at the end of our visit.

1858/01/02.	Dined at Buxtons.
1858/01/04.	FG all day at the models.
1858/01/11.	['Reports from the Niger Expedition' read before the RGS by Dr. William Baikie, FG asks a question.]
1858/01/14.	George made Vice Principal.
1858/01/23.	To opera. Prince of Prussia.[85]
1858/01/28.	Jules Gerard to dinner.
1858/02/25.	E. Darwin to Dinner.[86]
1858/02/05.	FG dined Mr Hirst.[87]
1858/03/02.	Tom Taylor to dinner.
1858/03/15.	Eclipse faded (FG at Oxford).[88]
1858/03/23.	To Hesketh Biggs.
1858/03/25.	[Emma] Frank and Louisa & me.[89]
1858/03/31.	FG missed train by going to Swindon.
1858/04/09	Mrs Phillips death.[90]
1858/04/10.	Darwin's offer to mother for Claverdon.
1858/04/11.	Home.
1858/04/12.	[FG and Louisa part from Emma][91]

[85] Three Princes of Prussia were visiting London: Frederick Charles, Frederick Albert and Adalbert.

[86] Erasmus Alvey Darwin (1804–1881), brother of the naturalist.

[87] Possibly Thomas Archer Hirst, a mathematician who was later an X-Club member.

[88] Solar eclipse at 12:05, which lasted only 2 seconds. The central line passed through Steeple Aston, north of Oxford. The view was spoiled by clouds.

[89] Emma's Diary.

[90] Galton's brother Darwin had married Mary Elizabeth Phillips (1822–1869).

[91] Emma's Diary.

1858/05/06.	Albert Smith was lecturing.[92]
1858/05/10.	[Review of *Western Africa* by Hutchinson in PRGS]
1858/05/29.	Box Hill.
1858/05/31	Home.
1858/06/01.	[Emma] To London, F. and L. to June.[93]
1858/06/02.	Cyril Graham to dinner.
1858/06/15.	Jenkinsons to tea.
1858/07/10.	To Hadley.
1858/07/12.	[Letter from War Office re models donated.[94]]
1858/07/19.	Gayton.
1858/07/20.	Windermere. Then Anne Sykes.[95]
1858/07/29.	To the Hollins.[96]
1858/08/03.	Norths arrived & took rooms for a week.

92 Perhaps Albert Richard Smith (1816–1860), the Alpinist. Between 1854 and 1860 he gave a long-running series of lectures on his ascent of Mont Blanc in 1851. That July he left for a trip to Hong Kong.

93 Emma's Diary.

94 Galton 1858. Galton's models to illustrate the Arts of Camp Life were copied by the War Office. Another set was made by Galton himself and landed up in the collection of the South Kensington Museum. 'Their object was not only to interest and instruct individual soldiers, but rather to suggest to instructors the precise subjects on which practical classes in the arts of camp life may most usefully be engaged. A catalogue accompanies the case in which the models are contained, and in this an asterisk is placed opposite to those objects which Mr. Galton's experience leads him to prefer for the purpose. "An old campaigner's acquirements consist partly in knowledge and partly in handiness. Field lectures, illustrated by experiments, may convey the first to an intelligent novice, and it was hoped that these models might serve to explain what kind of things must be made by his hands, before he can acquire the latter." The examples illustrate the various modes of the production of fire, the procuring, purifying, and carriage of water, cooking, the uses of portable food, substitutes for boats, cattle enclosures, expedients for tools and appliances in various handicrafts, tenting, hutting, and various other needs of camp life.' *Catalogue of the Special Loan Collection of Scientific Apparatus at the South Kensington Museum* (1876).

95 Anne Elizabeth Broadley.

96 A house in the Lake District, at Grasmere.

1858/08/07.	Hawkins[97] & FG slept out at Bow Fell.
1858/08/10.	FG & Hawkins slept out.
1858/08/11.	F. slept out.
1858/08/18.	Left Hollins for Patterdale.
1858/08/20.	To Greta Lodge.[98]
1858/08/28.	To Castle End Cottage.
1858/09/05.	Cath. died S. Leone.[99]
1858/09/10.	FG sent off his Review MS on Africa.[100] Emma had been with us.
1858/09/15.	[Letter from Henry Barth.]
1858/09/16.	FG with his diagrams.
1858/09/18.	To Windermere.
1858/09/20.	[British Association meets at Leeds.]
1858/09/21.	North Deighton[101] leaving FG at Leeds (B. Assoc.)
1858/09/25.	['Hand Heliostat' read by Galton on Saturday before the B.A. Section G (Mechanical Science).]
1858/09/28.	FG returned [from the B.A. at Leeds].
1858/09/29.	W. Malvern.
1858/10/07.	Leamington.
1858/10/07.	[With Emma, October 7th to 13.][102]
1858/10/15.	['Hand Heliostat' published in *The Engineer*.]
1858/11/11.	Col Le Froy to dinner.[103]
1858/11/30.	F quite unwell giddy—sick.

97 Francis Vaughan Hawkins.
98 In Keswick.
99 Catherine Bowen, Louisa's sister.
100 Not traced.
101 Yorkshire.
102 Emma's Diary.
103 Lieut.-Colonel Sir John Henry Lefroy (1817–1890), known for studies of magnetism.

1858/12/11.	[Letter from Charles Sturt, the explorer of Australia.]	
1858/12/20.	To Edstone—then Claverdon.	

1859

[Louisa's Summary] Returned home early in January. Music lessons of Schachner. Visited Adèle at Bognor April 12. Frank joined Monty, Johnson & Brodrick in a walking tour.[104] Met Frank on the 21st at Portsmouth & went on to the Isle of Wight, also Monty & met Mamma and the sisters. Mr. Spottiswoode came down. We returned May 2nd. Frank taken ill at once & dangerously ill for a fortnight attended by Dr. Hodgson & Dr. Todd. Abscess at the back of the tongue, broke May 15. Emma came in June. Went to Stanmore, to the Cautleys & to the Hollins Grasmere for 6 weeks, Monty Mamma & sisters at Keswick. Returned to London in September & left on the 6th for Bonn, lived in Pension with the Von Salomons for a month. The James Andersons in the house. Met the sisters at Paris in October & remained a fortnight. Home in London on the 25th. Heavy gales. Royal Charter lost.[105] Montague elected head master of Harrow Nov 16, strong party for Barry, supported by Lord Aberdeen. Spent Christmas and the New Year at Leamington and Claverdon.

1859/01/11.	Benthams & Spottiswoodes to dinner called on Brookfields—Atkinsons frequently.	London.
1859/01/24.	[Remarks on Burton and Speke, PRGS.]	London.
1859/02/–.	[Experiments on Tea Making.]	London.
1859/02/01.	Dined at Jenkinsons.	London.
1859/02/04.	Owen's lecture on the gorilla.[106]	London.
1859/02/09.	[Letter from Richard Owen re the Gorilla's range.]	London.

[104] H. Montagu Butler, H. Vaughan Johnson and George Charles Brodrick.

[105] On 1859/10/26 the ship *The Royal Charter* was lost off Anglesey, along with some 200 other ships in the storm path.

[106] Friday evening at the Royal Institution.

1859/02/—.	[Frank on walking tour (in Cornwall?) with H.M. Butler, H. Vaughan Johnson and George Brodrick.]	
1859/02/21.	[Frank and Louisa meet at Portsmouth.]	Portsmouth.
1859/02/23.	Dined Eras Darwin. Brodrick to breakfast.[107]	London.
1859/03/25.	[Letter from Sir George Everest re height estimation.]	London.
1859/03/29.	Oswell dined.[108]	London.
1859/04/12.	[Adèle at Bognor.]	Bognor Regis.
1859/04/26.	Isle of Wight. FG & Monty called on Tennison.[109]	Isle of Wight.
1859/04/29.	FG & Spott[iswoode] went out mapping. FG bad chest.	Isle of Wight.
1859/05/03.	[Emma] My Mother and I and Lloyd went to Rutland Gate; we found Frank very ill.[110]	London.
1859/05/05.	FG very ill in bed.	London.
1859/05/05.	[Emma] Left as Frank was ill and went to the Hotel, Crystal Palace.[111]	London.
1859/05/07.	Dr Hodgson came in evening.	London.
1859/05/10.	Dr Todd came Champagne & brandy every 2 hour.	London.
1859/05/12.	FG certainly better.	London.
1859/05/16.	Coughing, great relief. Dr Hodgson's 8th visit.	London.
1859/05/21.	Hodgsons farewell visit, puts stop to Sinai.	London.
1859/05/24.	FG drove out for an hour. Miss [Keary] to tea. Burton called for nearly 4 hours.	

107 George Charles Brodrick.
108 William Cotton Oswell.
109 Tennyson.
110 Emma's Diary.
111 Emma's Diary.

1859/05/27.	[Emma] To Rutland Gate till 20th June.[112]	London.
1859/06/13.	[Remarks on Burton and Speke.]	London
1859/06/16.	Grasmere Mrs Lomax. All the Butlers there—to Miss Parkers at Dudson Hall.[113]	Lake District.
1859/06/26.	[Letter from Charles Sturt re water.]	Lake District.
1859/08/06.	[Letter from Richard Burton about travel plans.]	Lake District.
1859/09/06.	[Left London for Bonn.]	
1859/09/09.	Settled at the V. Salomons. Called on Capt. Washington. FG went to Oolberg (excursion). I was busy all this summer on *Art of Travel.* FG & Jas. Anderson.	Germany.[114]
1859/10/02.	Dined at the [Buxtons].	
1859/10/11.	Left Bonn [...] Paris	France.
1859/10/25.	Left Paris for home—Uncle Henry to Devonshire Terr.	London.
1859/11/11.	To Stanmore Friday—Monday Miss Waddels.	
1859/11/15.	Montagu's election.	
1859/11/16.	To see the Arctic relics.[115]	London.
1859/11/18.	[FG at meeting of Marleybone Volunteer Rifle Corps.][116]	London
1859/11/28.	[Hand Heliostat, PRGS.]	London
1859/11/30.	Bristowes to dinner.	London.
1859/12/09.	[Letter from FG to Charles Darwin regarding *Origin of Species.*]	
1859/12/10.	[*Art of Travel,* Third Edition (Stated date, 1860).]	

[112] Emma's Diary.

[113] Possibly the illegitimate descendants of Galton's grandfather, Dr. Erasmus Darwin.

[114] Thus missing the meeting of the British Association, held that year at Aberdeen.

[115] Of the Franklin expedition, discovered by McClintock.

[116] *London Evening Standard,* Saturday 19 November 1859, 3.

1859/12/13.	[Letter from Charles Darwin regarding Origin of Species.]	Claverdon.
1859/12/29.	Macaulay's death.	Claverdon.

1860

[Louisa's Summary] Went to Leamington & Claverdon early in January. Let our house from May 1st for two months. Took lodgings at Richmond. Constant bad weather. Home sick to a painful degree & Frank almost always away in London or at Kew Observatory. Frank went to the British Association at Oxford end of June & I to Harrow 7 staid over the speeches. Frank sailed in the Himalaya July 6th for Bilboa, saw the eclipse from La Guardia. I joined him at Bordeaux July 25th. Travelled on together through the Pyrenees, staying at Luchon chiefly. Took Grenoble on our return. Back October 13th spent a few days at Leamington & Claverdon, home the 25th. Visited Monty at Harrow.

1860/01/08.	FG & Monty to Sevenoaks.[117]	
1860/01/09.	F. to Macaulay's funeral. Leamington. Edstone.	
1860/02/22.	Speke to breakfast.	London.
1860/02/29.	C. North to stay.	London.
1860/03/30.	FG at work on Speke's paper.[118]	London.
1860/03/31.	FG to Oxford with Spottiswoode.	Oxford.
1860/04/01.	[42 Rutland Gate is let for 2 months.]	
1860/04/04.	[Letter from Charles Frederick Frazier Mackenzie.]	
1860/04/17.	Miss Saas.[119]	London.
1860/04/23.	[Aunt Susan's] funeral.	
1860/04/30.	[Letter from Andrew Ramsay.]	London.
1860/05/01.	To Richmond to stay & Kew.	London.
1860/05/13.	[Letter from Charles Darwin.]	London.

117 A town in Kent, south-east of London.
118 See Speke 1860.
119 Possibly Florence Saas, the wife of the explorer Samuel Baker.

1860/05/27.	FG to [Airy[120]].	
1860/06/07.	[FG elected Fellow of the Royal Society.]	London.
1860/06/15.	[Letter from Sir Benjamin Collins Brodie.]	London.
1860/06/24.	FG giddy.	
1860/06/27.	[FG at British Association in Oxford. Elected to the council for 1860–1.]	Oxford.
1860/06/25.	[Louisa?] Left Richmond.	
1860/07/03.	[Emma] July 3rd to 9th, Rutland Gate.[121]	London.
1860/07/06.	FG left [for Spain, on the Himalaya].[122]	Spain.
1860/07/09.	[FG arrives at Bilbao.]	Spain.
1860/07/18.	Day of eclipse. [With Atwood and Gray at La Guardia, near Logroño.]	Spain.
1860/07/19.	[Letter from Charles Andersson, Cape Town.]	Spain.
1860/07/—.	[FG is in St. Sebastian.]	Spain
1860/07/25.	[Louisa] Met FG at station. Bordeaux.	France.
1860/07/28.	[Pica] Eaux-Bonnes [Bains de Bigorre].	France.
1860/08/06.	Luchon. Ch. Packe & C. Gray.[123]	France.
1860/08/21.	[Ascended Mount Néthoud.[124]]	
1860/08/25.	They & I up to sleep with the Russells.	

[120] George Biddell Airy.

[121] Emma's Diary.

[122] George Airy recorded the date as the 7th.

[123] Charles Packe (1826–1896) and Charles Gray.

[124] 'Charles Packe (England), Francis Galton, Charles Gray, Ellis Cunliffe. Our guide was P. Barrau. Left the Rencluse at 3 pm 45, arrived at the top in 3 h. 45. Mr. Ch. Packe left at a minimum thermometer, marking at 8 o'clock. 3 a.m., –3 centig. Barometer marked, 19.80 inches. Wind N. W. A fine day, but strata of clouds over the lower summits. Stayed 1 hour at the top.' Gray went up on the 28th again without Galton. *Revue des Pyrénées*, Volume 5 edited by Julien Sacaze, Felix Garrigou.

1860/09/10.	Left Luchon. Gavarnie Cauterets Toulouse Perpignan Canigou [Nimes] Uriage les Bains Lyons. Paris.	France.
1860/10/13.	Home. Leamington. Rugby [football].	London.
1860/10/20.	Claverdon.	
1860/10/25.	[Back in Rutland Gate.]	London.
1860/11/09.	Dined at Guildhall with R. Gurneys.	London.
1860/12/16.	FG fever.[125]	

1861

[Louisa's Summary] Easter at Hastings. Montagu ill at Tours. Summer tour to Switzerland. Met the Russell Gurneys at Champery, Frank to the Eggisch,[126] and both of us to Zermatt over Monte Moro to Italy. Heat intense. Returned by St Gotthard spent a fortnight at Bonn. Ill all the time with stiff neck. Visited Leamington & Claverdon. Montagu married Dec 19. Prince Albert died Dec 14. Very bad sore throat at Xmas, prevented going to the Boultons.

1861/01/04.	R Gurneys, fairy tale by G Macdonald.	
1861/01/11.	Hastings.	Hastings.
1861/01/15.	Ct. Blackiston [sic][127] and Oswell.	
1861/01/28.	[Additional instrumental instructions to Mr. Consul Petherick, PRGS]	London.
1861/02/04.	Lecture at SPG on Zanzibar.[128]	London.
1861/02/15.	Called on Kearys.	

[125] FG diary: Atwood. Bowden.

[126] Eggisch-horn.

[127] Captain Thomas Wright Blakiston (1832–1891) of the Royal Artillery. The following year he won the Founder's Medal of the RGS for his explorations of China, on which he was about to depart when he met Galton.

[128] Galton 1861d.

1861/02/25.	Geograph. about the Gorilla.[129]	London.
1861/03/03.	Sun in drawing room.	
1861/03/15.	[*Vacation Tourists and Notes of Travel in 1860* published.]	
1861/03/28.	Hastings about 4.	Hastings.
1861/03/31.	[Easter walking tour of Cornwall with Tyndall and Hawkins.]	Cornwall.
1861/04/11.	[Emma] Darwin [Galton]'s attack Thursday April 11th to May 30th; 7 weeks.[130]	
1861/04/17.	[Emma] [Darwin Galton] epileptic attack.[131]	
1861/04/21.	['On a new principle for the protection of riflemen', in *Journal of the Royal United Services Institute.*[132]]	
1861/04/29.	Bristed dined.[133]	London.
1861/05/—.	Colonial exhibition.	
1861/05/14.	We both dined at the Lord Chancellor's (Campbell).	
1861/05/21.	['African Exploration', in PRGS.[134]]	London.
1861/06/03.	Horticultural gardens opened (much seen of the Norths and Atkinsons).	
1861/06/29.	At Harrow. Great fire in London.	Harrow.
1861/07/—.	['Meteorological charts' in *Philosophical Magazine.*]	
1861/07/—.	*Circular Letter to Meteorological Observers. Synchronous Weather Charts* (Pamphlet).	
1861/07/24.	Paris Macon, Geneva, Champery Zermatt.	Switzerland.

[129] Paul du Chaillu had caused some controversy through his anthropomorphic reports from West Africa of the gorilla. See Galton 1861b.

[130] Emma's Diary.

[131] Emma's Diary.

[132] Galton 1861c.

[133] Charles Astor Bristed.

[134] Galton 1861a.

1861/08/18.	FG for Monte Rosa (monsieur a la belle[135] etoile).	Switzerland.
1861/08/20.	Tyndall (FG with Gray).	Switzerland.
1861/08/26.	Mattmark Sea, over Moro (with Hawkins) Macunaga.[136]	Switzerland.
1861/08/30.	Baveno.[137]	Switzerland.
1861/09/03.	Lugano. F.G. to Generoso.	Switzerland.
1861/09/07	Airolo.	
1861/09/09.	St Gotthard to Brenner Lucerne Basle Mayence Cöln.	Switzerland.
1861/09/14.	Bonn FG writing his meteor. paper.	Germany.
1861/09/28.	Brussels.	Belgium.
1861/09/30.	Home.	London.
1861/10/05.	Claverdon to 11th, then Leamington.	
1861/10/14.	F.G. to see the gaol.	
1861/10/25.	Home.	London.
1861/10/29.	Heard of Monty's engagement.	
1861/11/05.	To see "poor Mrs. Atkinson" but brave.[138]	Switzerland.
1861/11/13.	Sabines Bentham Tyndall to dinner.	London.
1861/11/15.	Mrs. Atkinson came & W. Spott: to talk over, warm discussion.	London.
1861/12/—.	[*Synchronous Weather Chart* (Pamphlet).]	London.
1861/12/15.	Heard of Prince Albert's death.	London.

[135] Galton has written 'bette', but this cannot be right.

[136] Macugnaga. Francis Vaughan Hawkins.

[137] This year Galton missed the British Association, held at Manchester on September 4.

[138] Thomas Witlam Atkinson had died on Aug. 13. After his death, his bigamous marriage was discovered.

1862

[Louisa's Summary] Exhibition year.[139] Visited the Norths, at Easter. Went to Switzerland in Autumn. [Murren], [Champery], Mont Chalet, where met the R. Gurneys, then to [Charmounix] with Miss [B...]. Arthur began Haileybury College. Association at Cambridge.[140] Frank went; I to Julian Hill & after to Hadzor & Leamington. Christmas at Harrow.

1862/01/27.	[Remarks on Baron von der Decken's East African Expedition, PRGS.[141]]	London.
1862/02/05.	Pooles to dine.	London.
1862/02/11.	Dined at Frere's.[142]	London.
1862/02/13.	FG wrote for Cornhill (Australia).[143]	
1862/02/20.	FG to Lord Ashburton left 28.[144]	The Grange, Hertfordshire.
1862/03/—.	['Recent discoveries in Australia' published in *The Cornhill Magazine*.[145]]	
1862/03/03.	Cyril Graham.[146]	London.
1862/03/11.	Talk with E Gurney about the 9 children.[147]	
1862/04/—.	[Emma] April to June, Rutland Gate.[148]	London.
1862/04/17.	Hastings.	Hastings.

[139] Great London Exposition, May 1862 – November 1862, held in South Kensington.

[140] British Association for the Advancement of Science.

[141] Galton 1862a.

[142] Possibly Henry Bartle Frere (1815–1884).

[143] Galton 1862b.

[144] William Bingham Baring (1799–1864) the second Baron Ashburton. His second wife Louisa Caroline Stewart-Mackenzie (1827–1903) had a literary-artistic-scientific salon.

[145] Galton 1862b.

[146] Sir Cyril Clerke Graham (1834–1895) the Orientalist who had explored the upper Nile in 1857.

[147] Emilia Russell Gurney (1823–1896), a friend of Louisa. The Gurneys were related to the Galtons. See *Letters of Emelia Russell Gurney* (London: James Nisbet, 1902) ed. Ellen Mary Gurney.

[148] Emma's Diary.

	Miss Ewart.	
1862/04/20.	Home.	London.
1862/04/26.	[*Vacation Tourists and Notes of Travel in 1861* published by Macmillan.[149]]	London
1862/05/03.	Alpine club in eve. Opening of Exhibition V Johnson to talk about his marriage.[150] Poles &c.[...]	London.
1862/05/24.	Brodrick to breakfast.[151]	London.
1862/05/26.	['Report on African explorations' in PRGS.[152]]	
1862/06/06.	Tyndall's lecture on force.[153]	London.
1862/06/18.	Mr Cantley.[154] (continually at the exhibition).	London.
1862/07/22.	Newhaven Paris Basle Zurich Stachelberg.	Switzerland.
1862/08/02.	Pilatus—Thunder storm at night.	Switzerland.
1862/08/—.	Rosenlaui Reichenbach Engatlen Brienz Giessbach [R.G. Fanthorn].	Switzerland.
1862/09/16.	Thun. Murren (host) Simmenthal [...] Aigle Miss Brandram [Chamonix] F. & Miss B. Mev. de [...] F. & Miss B. many walks	Switzerland.
1862/09/26.	Home.	London.
1862/10/01.	[FG at the British Association in Cambridge.]	Cambridge.
1862/10/03.	Emily to Haileybury (Rev F Smith of Harrow) Hadzor Leamington.	

[149] Galton 1862d.
[150] H. Vaughan Johnson, a friend from Trinity College days. He was married to a sister of F. Campbell, who had been at Cambridge with them both.
[151] Possibly George Charles Brodrick (1831–1903).
[152] Galton 1862c.
[153] John Tyndall delivered a noted lecture on force at the Royal Institution on Friday June 6, 1862. See *Proceedings of the Royal Institution* vol. 3 (1862), pp. 527–535.
[154] Possibly Nathaniel Cantley, who was a member of the RGS.

1862/11.	Eyelid cut by Bowman.[155]	London.
1862/11/15.	Lord Dundreary.[156]	London.
1862/12/04.	Gave Ramage warning.[157]	London.
1862/12/11.	Tom Hughes dined.	London.
1862/12/18.	Pilkington to breakfast.[158]	London.

1863

[Louisa's Summary] Easter at Leamington. Fell ill soon after with low fever, moved to Harrow June 5th, Frank had ague. He gave up being secretary at the Geographical after repeated worries about Dr Shaw. We both went to Brighton and Eastbourne in July. Frank spent August in Switzerland, I being at Julian Hill. He returned for the Association at Newcastle, after this we went to Switzerland. Grindelwald, Glion, over the Simplon & on to Florence with Spencer & Mary just after their marriage. Returned by the Riviera road & home in November. Frank appointed secretary to the British Association. Wrote paper on Domestication. Visited the Norths. Christmas at Julian Hill. Uncle Oxenham died in September.[159]

1863/01/02.	To Norths, Hastings Lodge.	Hastings.
1863/01/05.	Hinchcliffe & Parish.[160] at dinner.	Hastings.
1863/01/19.	"Reader" out.	London.
1863/02/04.	Mr Cator.[161] to dine. Coutts Trotter[162] & Jules Gerard to breakfast.	London.

[155] William Bowman (1816-1892) the ophthalmologist, and old associate of Galton from his days as a medical student in Birmingham.

[156] A character created by Tom Taylor. Perhaps the Galtons saw the play, *Our American Cousin*.

[157] One of Louisa's many evanescent servants.

[158] Possibly James Pilkington (1804-1890), who was a member of the RGS and an MP.

[159] William Oxenham (1800–1863), a master at Harrow, who married Louisa's maternal aunt Rachel Gray.

[160] Thomas Woodbine Hinchcliffe (1825–1882), the Alpinist and explorer, who had journeyed in South America. Sir Woodbine Parish, who lived at St Leonards.

[161] This may be C.O.F. Cator, who invented an anemometer.

[162] Maternal grandfather of J.B.S. Haldane.

1863/03/10.	Illuminations.	London.
1863/03/12.	Dined at Wickhams.[163]	London
1863/03/23.	FG decided to give up his secy R.G.S.	London
1863/04/04.	Claverdon.	
1863/04/11.	Home, 3 hours at Gayton.	London.
1863/04/18.	F.G. gave me a Davenport.[164]	London.
1863/04/27.	Heard of Mr Maitland's death.[165]	London.
1863/05/—.	[Emma] May to June, Rutland Gate, other visits till end of October.[166]	London.
1863/05/02.	FG went to Mr Lubbock's.	
1863/05/12.	To June 8 (at least) ill. [FG] She must have been at home.	London.
1863/06/03.	She went to Harrow & FG too.	Harrow.
1863/06/14.	FG complained of illnesses. He went to London.	
1863/06/19.	Decided account if RGS argue.	
1863/06/25.	FG returned unexpectedly, looking ghastly drove with him to the station.	
1863/06/30.	Went together to Brighton.	
1863/07/13.	[Emma] Unprotected came out!![167]	
1863/07/21.	Returned home. Saw Mary Kendall & was pleased with her. F.G. went abroad.	
1863/08/11.	Heard from FG at Grindelwald (? avalanches).	Switzerland.
1863/08/24.	F.G. returned [for B.A. at Newcastle].	London.
1863/09/04.	Back home (from Harrow).	London.

[163] Possibly William Wickham (1831–1897).

[164] Sofa.

[165] Possibly John Gorham Maitland (1818–1863), who had been at Cambridge shortly before Galton. He was the father of the historian F. W. Maitland.

[166] Emma's Diary.

[167] Emma's Diary. Her book *A Guide for the Unprotected.*

1863/09/12.	Folkestone; Boulogne; Paris; Dijon; Dole; Lausanne; Triberg; Berne; Thun; Glion met Spencer & Mary; Brienz; Simplon; Domo a' Opera Florence; Pisa; Spazzi; Sestri; Genoa; Suse; Mt. Cenis; Geneva; Paris.	
1863/11/10.	Home.	London.
1863/11/12.	F.G. to Mr Lubbock.	London.
1863/12/09.	Helped F.G. about the Handbook.[168]	London.

1864

[Louisa's Summary] Very cold beginning of the year. Went to Leamington at Easter. Got sore throat. Emma & Milly came to us in May. Went abroad in July to Switzerland. Peiden Grindelwald, St Luc. Returned for the Association at Bath. I at Julian Hill meanwhile. Visited Hadzor & Leamington. Fell ill soon after. Went to the Norths in December. Frank writing Hereditary Talent. Agnes born. Arthur fell ill from overwork. Eva killed August 18th. Cyril born in July. Frank busy editing Murray's handbook Christmas at home & alone.

1864/01/05	dined at Aytouns	London
1864/01/06	Mrs Atkinson and Alatau[169] to dinner	London
1864/01/21	saw Mrs. Boulton's death	London
1864/01/26	FG at Ethnological (Had been attending Tyndall's lectures)	London
1864/02/—	[Emma] February to March[170]	
1864/02/03	Mr Prideaux called[171]	London

[168] Of Switzerland.

[169] The widow of Thomas Witham Atkinson, the Siberian traveller, and their son Alatau Tamchiboulac Atkinson. T.W. Atkinson was a bigamist, and his second 'marriage' was found invalid after his death. See Galton 1908b, 176–177.

[170] Emma's Diary.

[171] Lieut. William Francis Prideaux (1840–1914), who had been in the Indian Army and later acted as consul at Zanzibar. He retired as a full colonel and published literary scholarship.

1864/02/09	L. Gurney took me to Ethnological in Evg	London
1864/02/10	FG dined at the J. Kay-Shuttleworth[172]	London
1864/02/17	dined at Markhams	London
1864/02/22	at Douglas' Prestwich & Falconer[173]	
1864/03/08	Ethnological, Mr. Lubbock's paper	London
1864/03/19	Leamington	
1864/03/28	Claverdon	
1864/03/31	Leam.	Leamington
1864/04/02	Home	
1864/04/11	[Galton attends Garibaldi's procession through London (freedom of the City) at Trafalgar Square]	
1864/04/26	Bates[174] lecture at the Horticultural	
1864/05/—	[Emma] May to June, Rutland Gate.[175]	
1864/05/01	FG to Lubbocks	
1864/06/21	FG at Ethnological	
1864/07/05	Stanmore	
1864/07/11	home	
1864/07/15	Dieppe; Basle; Schaffhausen; Appenzoll; St Gau; Robach; Pfaffers; FG did his index; Dossentis; Furka; Handek; Imhof; Grindelwald; Spitz; Kandersteg]; Gemmi; St Luc.; quite taken ill (George B's child over bannisters & killed)	Switzerland

[172] Sir James Kay-Shuttleworth (1804–1877) the elder brother of Joseph and Eben Kay, who were at Trinity College Cambridge with Galton. He married Janet Shuttleworth.

[173] The geologists Sir Joseph Prestwich (1812–1896) and Hugh Falconer (1808–1865).

[174] Henry Walter Bates.

[175] Emma's Diary.

1864/08/20	Bella Fola	
1864/08/22	FG to Zurich & met Les. Stephen; Aisle; Lopez; Comballia; H. Byron	Switzerland
1864/08/27	Paris	France
1864/09/07	Home	London
1864/09/10	Harrow	
1864/09/15	[FG at B.A. Bath] with Lubbock & [...]	Bath
1864/09/22	FG back (?from where[176]) Leam.	
1864/09/29	Consecration of Miss Ryland's Church	
1864/10/01	heard of Uncle Hubert's death	
1864/10/04	Leamington. FG had been to Cheltenham to see George	
1864/10/10	Hadzor	
1864/10/18	Home.	London
1864/10/22	Harrow	
1864/10/28?	Mr Lubbock &c. to dine	London
1864/11/29	Hastings Lodge. Sir W[oodbine] Parish.	
1864/12/02	FG to Rye	
1864/12/05	Home	London
1864/12/06	Copied FG's Reader article[177]	London
1864/12/14	[...] to dinner	London
1864/12/21	Inskip (the maid) left Elliot came	London
1864/12/27	Did weather charts with FG	London

1865

[Louisa's Summary] Fell ill Jan 7th, moved to Harrow in March. We both went to Hastings & St Leonards at Easter. Better after this, but not strong. Ordered to Spa. Frank went to Wales & Dublin first, had bad weather. He published two papers on Hereditary Talent in Macmillan, June & August. Reviewed Frost & Fire in Edinburgh Review. British Association at

[176] B.A. at Bath.

[177] Likely "Burton on the Nile Sources" *The Reader* 10 December 1864.

Birmingham in September. Visited Adèle. Claverdon Hadzor & Leamington. Frank fell ill at Hadzor with cold, followed by boils, which troubled him much through winter. Engaged on the part of the Royal Society to look over the meteorology as left by Fitzroy. Xmas at Harrow. Norths in Egypt. General election in July of this year.

1865/01.	Weekly Reader meetings, Tuesdays at Tom Hughes'.	London
1865/01/09.	Ld Campbell in Evening. FG was at Ethnolog. every Tuesday.	London
1865/02.	Letters about Hopkins' failure of intellect.	London?
1865/02/11.	began to arrange Brit. Assoc. What is the "Century".	London?
1865/03/06.	LG had been v. unwell - Harrow.	Harrow
1865/03/24.	Home.	London
1865/04/03.	Hastings 34 Marina Symmonds (? is C.N. married, yes.) Hastings Lodge.	Hastings
1865/04/30.	[Admiral FitzRoy commits suicide]	Hastings
1865/05/01.	Left Hastings.	
1865/05/08.	[Emma] May 8 to 22 Rutland Gate.[178]	
1865/05/13.	FG to Newstead Abbey to 15th.	
1865/05/31.	FG to the Derby drive of 3 hours.	?
1865/06/15.	Early to De Mussy (read Palgrave).	London
1865/07/04.	[FG] exped. by self to Wales.	
1865/07/06.	dismissed Skeat.	London
1865/07/11.	[FG] 1 day at Dublin.	
1865/07/17.	[FG] home.	
1865/07/20.	Calais Brussels Spa (Dr Lizack[?])	Belgium
1865/08/16.	Mr Papillon ill.	
1865/08/16.	Liege on Steamer, busy day.	

[178] Emma's Diary.

1865/08/18.	fright about him [Papillon]. Crosswells.	
1865/08/21.	Dutch tour to 24th Roy. Soc. Council.	
1865/08/24.	left Spa, Brussels.	Belgium
1865/08/28.	Home.	London
1865/09/04.	Macgregor (Rob Roy).	
1865/09/05.	Brit Assoc Birmingham Phillips Pres. (we were at Evans[179]) [Pres] Forbes.	Midlands
1865/09/09.	Oaklands, lunch with Matthews (Alpine maps) Whymper.	
1865/09/18.	Leamington Claverdon.	
1865/09/25.	Hadzor Mr Temple (?diving specs).	
1865/10/13.	Home.	London
1865/10/19.	Haileybury[180]	
1865/10/24.	Home (Mr Venn).	
1865/10/30.	FG boil.	
1865/11/13.	Baker's paper at the Geogr.[181]	London
1865/11/25.	[...] at [Manchester] to 27th.	
1865/12/05.	FG quite well at last.	London
1865/12/09.	Agnes Zimmerman 17th lesson.	London
1865/12/14.	FG had a hand in Report on Meteorology finished 15th.	
1865/12/19.	Gurneys fixed to go [to] Jamaica.	
1865/12/23.	To Harrow.	

1866

[Louisa's Summary] Russell Gurneys off to Jamaica. Frank regularly knocked up in July & unable to finish his meteorol-

---

[179] Possibly John Evans, who had been accompanied by Galton and Lubbock when investigating Stone flints and other implements at Little Salisbury Hill in 1864, during the nearby meeting of the B.A. at Bath. See *Transactions of the Ethnological Society* 4 (1866): 240.

[180] Arthur Butler.

[181] "Account of the Discovery of the second Great Lake of the Nile, Albert Nyanza." By Samuel White Baker, Esq. Read, November 13, 1865.

ogy. Went to Brighton in March for a week, then abroad to Cannes, Menton, Genoa, Italian lakes & home by the Simplon to Glion. Met Emma at Boulogne in June reached London on the 20th, our house let. Paid visits at Julian Hill & Grasmere, staid there 6 weeks. Rain constant. Emily & Gertrude & Montagu joined us. The end of August went to Nottingham for British Association meeting, Frank drove up & obliged to leave. Went straight to Bertie Terrace. Adèle lent us her house Edgemead. Dr. Jephson prescribed for Frank, he grew very weak under the treatment. End of September returned home & remained six weeks, then went to the Norths & took lodgings at Hastings in Breeds Place, staid there till near end of Jany 1867. Frank rode constantly. Montagu's boy Teddy born in December.

1866/01/10	Bad night from FG's cough.	London
1866/01/18	Dined with Reeves, (C[ommander]) Evans 16-19 (Tyndall is lecturing).	London
1866/02/07	FG Poorly & giddy. (My regular breakdown).	London
1866/02/10	[De Mussy.[182]] ordered us abroad	London
1866/02/11	FG worse by talking to Farrer & Grove.	London
1866/02/14	[Violetta and Emma Sophia Galton write to urge rest.]	London
1866/02/20	FG very poorly after talk with Sabine & Murchison.	London
1866/02/23	[Letter from Murchison.[183]]	London
1866/03/03	[Emma] Frank ill, overdone with volunteer work and must go abroad.[184]	
1866/03/03	Boulogne.	France
1866/03/08	Arrived Villa Allegria at Cannes (lent us by Mrs Hollond).	France
1866/03/15	to a villa in Cannes.	France
1866/03/26	Heard of Camerons death.	France

[182] Their physician, see Register.
[183] GALTON/3/2/1/95
[184] Emma's Diary.

1866/03/30	FG lost his plaid in the [Escalles].	France
1866/04/04	Grasse & back.	France
1866/04/07	Menton.	France
1866/04/20	FG walk with Mr [Moggridge].	France
1866/04/23	Bordighera.	Italy
1866/04/24	Alassio.	Italy
1866/04/26	Genoa (the Ottleys) Milan.	Italy
1866/05/03	Cadenabbia FG & Hawkins to Lecco.	Italy
1866/05/15	Lugano.	Italy
1866/05/17	Jenkinsons came. [Symonds].	Italy
1866/05/23.	Stresa (my house was let).	Italy
1866/05/28	FG & Mr Hawkins to Orta.	Italy
1866/06/01	Baveno.	Italy
1866/06/04	Simplon to Brieg.	Switzerland
1866/06/06	Glion.	Switzerland
1866/06/13	Neuchald.	Switzerland
1866/06/14	Paris.	France
1866/06/16	Boulogne. Emma arrived but left 18th.	France
1866/06/20	[Boulogne. FG observes meteor there.[185]]	Boulogne
1866/07/05	Haileybury.	England
1866/07/17	Peterborough.	England
1866/07/19	Grasmere.	Cumbria
1866/08/01	to Ambleside & back [saw dykes].	Cumbria
1866/08/13	Called on the Halle's.	Cumbria
1866/08/13	[Letter from Buys Ballot about Maury & weather prediction.[186]]	Cumbria
1866/08/14	[LG] did diagrams with F.G. Montagu came, Temples.	Cumbria
1866/08/19	F & M[ontagu] to Ambleside to dine with the Temples.	Cumbria

[185] BA report 1866, p. 126

[186] GALTON/3/2/1/3. Christophorus Henricus Diedericus Buys Ballot (1817-1890) was a Dutch meteorologist.

1866/08/19	[Letter from Archibald Smith about Wind Charts.[187]]	Cumbria
1866/08/22	FG giddy. Brit Assoc Leeds.[188] FG done up. Ramsay in carriage.	Nottingham
1866/08/24	[To Mother at Leamington.]	Leamington
1866/09/03	FG poorly.	Leamington
1866/09/15	Jephson put FG on strict diet [The ... at Kavanaugh].	Leamington
1866/09/28	Home.	London
1866/10/09	Griffiths came on Association business.[189]	London
1866/10/10	[Letter from F. Plateau dissenting on 'Spectacles for Divers'.[190]]	London
1866/10/11	FG quite tired.	London
1866/11/10	FG saw de Mussy (Genl Sabine was here).	London
1866/11/15	[To Hastings.]	Hastings
1866/11/26	FG rode for 4 hours (? Miss Br[andram]).	Hastings
1866/12/30	[Letter from JA Symonds re pedigrees of Venetian painters.[191]]	Hastings

Notes: Galton made notes on 'The Danger of the Deterioration of Race from the Too Rapid Increase of Great Cities' by John Edward Morgan.[192]

1867

[Louisa's Summary] Went to Italy end of Jany. Miss Brandram with us. Staid chiefly in Rome, Naples & the neighbourhood & Venice, then by the S. Tyrol to St. Moritz, where the cure did wonders for me, but did not suit Frank. Next

[187] GALTON/3/2/2/39
[188] An error, possibly made by FG when transcribing his wife's diaries. It was actually held at Nottingham.
[189] George Griffiths.
[190] GALTON/2/1/3/4.
[191] GALTON/2/4/1/1/27
[192] GALTON/2/12/28

to [Seelisberg], disappointed of Arthur through Montagu's illness. Then to Heidelberg, through Bavaria & home in October to Dover first, then to Julian Hill. Bertie Terrace, Malvern and Hadzor. Settled in London by end of November. Reform Bill passed in summer. Fenians in revolt. Frightful catastrophe at St Thomas. Eruption of Vesuvius in December. Amy Biggs died Aug 12 & Milly's 1st child born same day. Arthur resigned Haileybury in Dec. Montagu went to Cannes for Xmas holidays, Xmas at Julian Hill. Mother Mrs Galton gave me Astracan[sic] & Velvet cloak. Teddy dangerously ill.

1867/01/01	(Hastings) Latham. Sabines had just left.	Hastings
1867/01/04	Miss B. fixed going with us to Italy. Blakiston [Brodrick].	Hastings
1867/01/07–1867/01/08	[illegible]ssea. Latham - Cowell - Cook Rev & Montagu. George & Josephine.	Hastings
1867/01/21	Lord Ashley offered to take home. Accepted.	Hastings
1867/01/23	Heard our house is let.	Hastings
1867/01/24	Home.	London
1867/01/26	Lunch Athen[aeum]. with Hurst Huxley Dr. Smith & Spotti[swoode].	London
1867/01/30	[Emma] Frank sailed for Boulogne.[193]	
1867/01/30	Inventory. Began tour.	London
1867/01/31	Paris with Miss B.	France
1867/02/02	Dijon (Ed. Wh.)	France
1867/02/04	Hyeres.	France
1867/02/07	Cannes to Mrs Hollond and on to Nice. Menton, Alassio, Oneglia.	France
1867/02/10	Savona, Genoa, saw "Affondatore".[194] Chiavari (little birds).	Italy

[193] Emma's Diary.

[194] The ironclad battleship, named 'Sinker', equipped with a ram and two 300-pounder guns. It was built in England for use against the

1867/02/15	Spezzia. Mr Somerville in no 9.	Italy
1867/02/18	Pisa.	Italy
1867/02/20	Perugia.	Italy
1867/02/22	Rome.	Italy
1867/02/26	Rome saw horses start [carmine] was much with W.G. Clark.	Italy
1867/03/04	Vesuvius flavour of sulphur.	Italy
1867/03/08	The Mosaics (Mrs. Skinner the American rich lady) W.G. Clark.	Italy
1867/03/30	Naples.	Italy
1867/04/06	Sorrento.	Italy
1867/04/12	Posted memoir to Kew on mechanism for coordinating meteor[ological] curves ([Meuricoffres] frequently in Naples).	Italy
1867/04/13	FG sent off paper to Kew. Amalfio.	Italy
1867/04/20	Naples.	Italy
1867/04/24	Ischia.	Italy
1867/05/04	Naples & Capri.	Italy
1867/05/08	Naples.	Italy
1867/05/10	Ancona Loretto Ravenna.	Italy
1867/05/14	Venice.	Italy
1867/05/27	Counted pigeons at St Mark's.	Italy
1867/05/28	Padua Verona.	Italy
1867/05/29	Riva.	Italy
1867/06/03	Botzen.	Italy
1867/06/05	Meran.	Italy
1867/06/07	Nauders [Tarasp]	Austria
1867/06/14	St Moritz.	Switzerland

Austrians, and delivered to the Italian navy in 1866. After sustaining some damage in an early engagement, it sank in a storm on August 8th. It had been raised in 1867 and was being rebuilt in La Spezzia when the Galtons saw it.

1867/06/19	FG giddy - poorly - finished his book (Rev Streatall). Giddy but about.	Switzerland
1867/06/21	Finished revise of Art of Travel.	Switzerland
1867/07/09	[Letter from Vambery about BA in Dundee.[195]]	Switzerland
1867/07/11	Le Prese[196]	Switzerland
1867/07/16	FG finished Art of Travel (Col Weston).	Switzerland
1867/07/20	Col. Weston to the limestone pyramids Mt. [Pi'adella][197].	Switzerland
1867/07/29	Coire (by Albula)	Switzerland
1867/07/30	Zurich	Switzerland
1867/08/01	FG to [Pile] diggings at Wetzikon	Switzerland
1867/08/02	[Seeligsberg] ([Mrs. Cobbe]).	Switzerland
1867/08/07	[Letter from Florence Nightingale re apparatus.[198]]	Switzerland
1867/08/10	Lushingtons came. Spottiswoode too. Pension Gerig. (FG up Rigi[199])	Switzerland
1867/08/19	Basle.	Switzerland
1867/08/20	Baden.[200]	Germany
1867/08/20	[Letter from Samuel White Baker about Galton's health, suggesting Egypt for recuperation]	Germany
1867/08/26	Carlsruhe.	Germany
1867/08/27	[Letter from Sir Stafford Northcote re Mansfield Parkyns.[201]]	Germany
1867/08/28	Heidelberg. Amy's death.[202]	Germany

[195] GALTON/3/1/1/47
[196] Near the Bernina Pass.
[197] Pisciadella
[198] GALTON/3/2/1/99
[199] A middling mountain on the shores of Lake Lucerne.
[200] Baden-Baden.
[201] GALTON/3/2/1/98
[202] Amy Biggs.

1867/09/04	[FG misses] B. Assoc begins at Dundee	Germany
1867/09/18	Stuttgart.	Germany
1867/09/20	Ulm & Augsberg.	Germany
1867/09/21	Munich (FG coins tinfoil).	Germany
1867/09/30	Ratisbon.	Germany
1867/10/04	Würtzburg. Frankfort. FG has his Index proof.	Germany
1867/10/07	Bonn.	Germany
1867/10/10	Brussels.	Belgium
1867/10/12	Dover.	England
1867/10/19	Home.	London
1867/10/21	Harrow.	England
1867/10/29	Leamington.	England
1867/11/09	Claverdon.	England
1867/11/12	FG to Berkswell (H. Watson[203])	England
1867/11/18	Malvern.	England
1867/11/22	Hadzor.	England
1867/11/28	London.	England
1867/12/03	Strike of cabs.	London
1867/12/20	Whymper to lunch.	London

Notes: Letter from Julia Hallam regarding history of the Hallam family, 1867.[204]

1868

[Louisa's Record] The year began by Frank being too poorly to dine at Mr. Crawfurds. Winter at home, but dined out very little. Miss Munro in London. Emma came to us in March. We went to Leamington April 7. George & Josephine came to us on our return. Maggie the Lethbridges & Whelers came after, in succession. Mr Crawfurd died May 12th. F. went to

[203] The mathematician H. W. Watson, who later collaborated with Galton on branching processes. This visit was likely connected with *Hereditary Genius* and Tripos examination marks. Galton may have met him through the Alpine Club. Watson's daughter Maud Edith Eleanor was the first Wimbledon Lady's Tennis Champion in 1884.

[204] GALTON/3/3/12/15

his funeral. Not well in July. The Norths asked us to Hastings Lodge, met the Knox's there. A lodging at St Leonard's for a week. Parish's very kind.[205] Then went to Harrow for a fortnight & on to Leamington. Very fine hot summer. Crossed to Dieppe September and on to the Auvergne, Geneva & Lyons. Constant rain. Home Oct 27. General elections in November. Mr North returned but petitioned against. My digestion all wrong. Went to De Mussy. Spent Xmas at the Norths. Benthams & Headlams there. Left on New Year's day. The Norths very depressed about the petition. Mr Boulton remarried.

1868/01/01	London. FG poorly.	London
1868/01/02	[Letter from John Crawfurd inviting Galton to dinner][206]	London
1868/01/16	Munro to dinner[207]	London
1868/01/31	Saw Lady Dalyell	London
1868/02/11	Hastings	Hastings
1868/02/17	At Meteor. FG had a small row with Gen. Sabine.	London
1868/03/—	[Emma] March, Rutland Gate.[208]	
1868/03/07	[FG attends Royal Society Soiree at Burlington House[209]]	London
1868/03/14	['At their meeting on March 14th, Mr. F. Galton, general secretary, informed the council [of the British Association] that considerations of health precluded him, to his sincere regret, from continuing to hold office.'[210]]	London
1868/03/19	FG to Harrow about Geograph. Prizes.	Harrow

[205] Sir Woodbine Parish, from the Ethnological Society.
[206] GALTON/3/2/2/11
[207] Miss Munro.
[208] Emma's Diary.
[209] *Morning Post*, Monday 09 March 1868: 5.
[210] *Morning Post*, Thursday 20 August 1868, p. 2.

1868/03/23	[FG is part of a British Association deputation to visit Disraeli at 10 Downing Street[211]]	London
1868/03/–	[Letter from Francis Turner Palgrave re pedigree.[212]]	
1868/04/07	To Leamington	Leamington
1868/04/11	FG rode to Berkswell	West Midlands
1868/04/17	[Letter from John Crawfurd[213]]	
1868/04/21	Home	London
1868/05/—	[Emma] May and June, Rutland Gate.[214]	
1868/05/12	Mr Crawfurd's death. [Letter from classical scholar H.A.J. Munro.[215]]	London
1868/05/18	Funeral	London
1868/05/25	[FG's proposal for Geography Prizes at schools is put to the RGS council[216]]	London
1868/05/28	Wickham Court	London
1868/06/16	FG to Claverdon & Loxton	
1868/07/10	C. Buxton's water party to Richmond	London
1868/07/23	Hastings Knox came	
1868/08/06	Harrow	
1868/08/15	FG to Waverley	
1868/08/17	London	London
1868/11/28	[Letter from Isaac Taylor regarding pedigree.[217]]	
1868/12/01	[Letter from Benjamin Hall Kennedy re pedigree.[218]]	

[211] *London Daily News*, 24th March, 1868: 5.
[212] GALTON/2/4/1/2/5
[213] GALTON/3/2/2/11
[214] Emma's Diary.
[215] Galton 3/2/1/93
[216] PRGS 38 (1868): x.
[217] GALTON/2/4/1/3/14
[218] GALTON/2/4/1/2/3

1868/12/02	[Letter from James Fitzjames Stephen on family pedigree.[219]]
1868/12/04	[Letter from Charles H Gregory re pedigree.[220]]

Notes: Visited Foundling Hospital.[221] Galton made notes on Darwin's *Variation of Animals and Plants under domestication.*[222]

1869

[Louisa's summary] My digestion very troublesome till June and a great hindrance to my doing much. Frank in good health & able to dine out again. Went to Bertie Terrace at Easter & was not the better for it. Emma came to us in March & June. Lucy Wheler in May. Started for the Tyrol & Bavaria, Venice and home by the Splugen in July, returned October 4, met the R. Gurneys & Mrs. Batten at Basle. Went to Julian Hill & to Bertie Terrace before settling down. Dear Mr. North died October 26 at Hastings. Frank's book "Hereditary Genius" published in November, not well reviewed, but liked by C. Darwin & men of note. He began his experiments in transfusion & became a member of the Zoological & Royal Institution. I attended Tyndall's lectures after Easter. Spent Christmas at home & alone. Darwin's wife died Nov 26 after a month's illness.[223]

[219] Galton 2/4/1/3/13
[220] GALTON/2/4/1/4/9
[221] GALTON/1/2/4/8
[222] GALTON/3/1/1/14
[223] That is, Darwin Galton's wife.

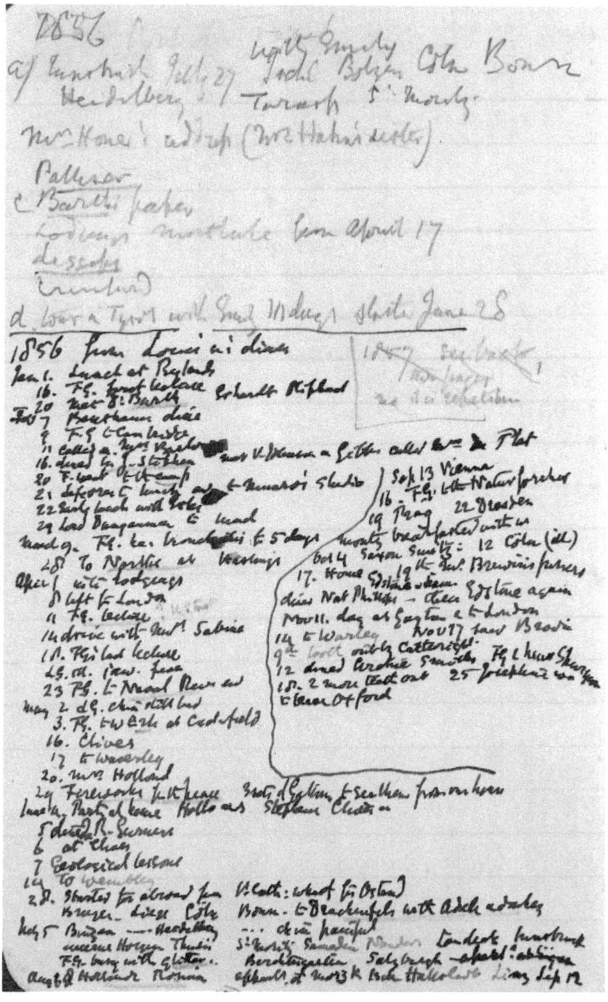

Figure 27.1: Page from FG's Memories *notebook, containing Louisa's Diary entries.*

Chapter 28

Louisa's Yearly Record

From Louisa Galton's Summary for each year.

1869

My digestion very troublesome till June and a great hindrance to my doing much. Frank in good health & able to dine out again. Went to Bertie Terrace at Easter & was not the better for it. Emma came to us in March & June. Lucy Wheler in May. Started for the Tyrol & Bavaria, Venice and home by the Splugen in July, returned October 4, met the R. Gurneys & Mrs. Batten at Basle. Went to Julian Hill & to Bertie Terrace before settling down. Dear Mr. North died October 26 at Hastings.[1] Frank's book "Hereditary Genius" published in November, not well reviewed, but liked by C. Darwin & men of note. He began his experiments in transfusion & became a member of the Zoological & Royal Institution. I attended Tyndall's lectures after Easter.[2] Spent Christmas at home & alone. Darwin's wife[3] died Nov 26 after a month's illness.

1870

Dear Mrs. Batten died Jany 26 and saddened our year. At Easter we went to Bertie Terrace & Mrs Galton was taken seriously ill & for 3 weeks we despaired of her life.[4] In May & June E. Wheler & Emma came to us. A long spell of

[1] North, Frederick. See page 360.
[2] Tyndall, John. See page 360.
[3] Galton, Darwin. See page 356.
[4] Galton, Frances Anne Violetta. See page 359.

fine dry weather beginning from Easter till October with little intermission. We went to Paris on July 15, intending to go to St Moritz, but the declaration of war on the 15th prevented that & though we went on to Grindelwald we returned in a fortnight & remained for a fortnight at Folkestone for sea bathing. Then we came home & visited the R. Gurneys, Julian Hill & Bertie Terrace, felt Mrs Galton's increased feebleness, but had a visit stored with nice memories. Went to the Groves at Arthingworth.[5] All this time the war raging, Napoleon capitulated in September after Sedan. Returned to London October 17. Our house painted & redecorated by Battam & Waller. Frank went to the British Association at Liverpool in October.[6] Both of us vaccinated end of June, I partially took in one place, Frank did not. Spent Xmas at home, extreme cold. Arthur less well.

1871

Went to Leamington at Easter & I to Julian hill after, while F. went with Montagu to Derbyshire. The Lethbridges, Emily & B. Gainsford, Lucy Biggs & Emma came to us between April & July. Trouble about Elliot & the cook, Mrs Cole. Elliot left us July 4. We went to Scarborough, Whitby, Yorkshire & Ilkley. The R. Gurneys to America about the Alabama claims. In September we went to Leamington & on to Julian Hill. I ill with a bad cold all the time. We left October 17 & went to Brighton for a week. Then home & not settled with servants till December. Frank gave up his rabbit breeding & took to machine inventing. A dull Xmas at home. The Montagus, George & Spencer came after. Mamma's last Xmas. She wrote sadly about it. Arthur at Cannes.

1872

A marked year in many ways & chiefly so by the death of my dearest mother on Feb 24 at Julian Hill after 6 days illness from a chill by going into the cold drawing room. She had no strength to bear up against the attack. The week before we were at Southsea enjoying the Dockyard at Portsmouth & the sight of the great ships of war. Capt Hall took us about in his

---

[5] Grove, Justice William Robert. See page 359.

[6] The British Association met at Liverpool on September 14-21, 1870.

steam launch. We went over the Wellington, the Victory, the St Vincent training ship, the Queen's Yacht, the Enchantress & the Monarch & Devastation the great ironclads, also the Trafalgar previous to its sailing next day. The Gassiots very kind to us. We were back home Feb 12 & had no uneasiness about dear mother till Feb. 23, the next day a telegram summoned us hastily, but she was not conscious when we arrived. She was buried at Harrow on Feb 29, a solemn but striking day to us all, the arrangements beautifully made by Montagu. I had not seen her since Oct 17, as she had been feeble & had servants changing & was not able to have us. All the months after till Augt., we were often at Julian Hill. We went to Leamington for Easter on March 27. Mrs. Galton well, left April 8. George & Josephine & sons came up in July. Family meeting at Julian Hill July 6. The Gurneys returned from America May 8. Many pleasant meetings with them. Frank taken up with Spiritualism and attended meetings at Mr. Crookes & Cox. We Went to Brighton on Aug 8 for the British Association, & Emma joined us on the 10th. Frank President of the Geographical Section. Stanley made himself most conspicuous & obnoxious. We left on the 24th & went home for a week & then Dr. de Mussy, on Sept 3. Very ill with cholera & F. with fever. Remained till Oct. 14, when hurried home on account of dear Emma's operation for ovarian tumour on Oct 17. All went on well & she came to us the 13th of December for a few days. We spent Xmas with the Douglas Galtons at Hadzor very pleasantly. A remarkably wet year. Frank busy writing on Hereditary subjects. A paper in Fraser. My sisters & Arthur went to Rome in December. Spencer to live at Julian Hill. Found the comfort of an extra servant this year. Missed the absence of E. Gurney & M. North very much. Archie Smith died Dec 26.

1873

The early part of the year uneventful. Darwin was married in Jan 7. We went to Leamington at Easter & found Emma very feeble. Lucy & the Lethbridges came to us in May. We left home early July 1st for [Thuringia] & did not extend our tour much on account of cholera, we staid at Bonn on our return. Frank came back for the British Association at Bradford & joined me at Leamington. The Gurneys & M A North returned in September. We settled at home the end of October

& had great bother about a cook. Emily settled at Harrow in the summer & Gertrude joined All Saints Sisterhood in July. Montagu's 2nd boy born & Spencer's 3rd boy. Spent Xmas at Julian Hill. A pleasant family gathering. Frank chiefly buried on the hereditary endowments of his scientific men.

1874

Uneasy from the very beginning about dear Mrs Galton, Frank went to see her early in Feby & she died Feby 12th aged 90. This coming so soon after my dear mother made a sad blank, both houses gone. We went to dear Emma at Easter for a week & then made a few days visit to Winchester Salisbury & Lyndhurst, the weather not good. At Whitsuntide we visited the Jenkinsons & greatly enjoyed their lovely country. George & Josephine came to us the end of June. We were prevented going away by domestic bothers, which have made the whole of this year sadly trying. Thompson left us in March & was a sad loss to me. We paid many visits during July & at the end went down to Cornwall spending a fortnight with Adèle & Milly & visited Boscastle & Tintagel. After this we went to N. Devon till September when we visited the Groves at Chettle near Blandford. On the 14th, I broke a blood vessel & was very near dying, but through God's mercy, I came back to life & felt so peaceful & happy in my quiet sick room, that it was not a time of misery. And all were so kind and good to me & Frank especially, that I felt sustained by love. We moved to Bournemouth as soon as I was able & then in November to dear Emma & found her well in her newly arranged house. We came home November 10th. Very severe weather soon set in & lasted to the end of the year. Frank was ill in December & had Dr. A Clarke.[7] We had a quiet dull Xmas, no goings out & F. had to give up his promised lectures at Newcastle. His book on Nature & Nurture of Scientific Men came out in December, occupied on enquiries about Twins. On the whole a year of sad memories. The dear Duchess died in March about a month after dear Mrs. Galton.

[7] Dr., later Sir, Andrew Clarke (1826–1893).

1875

Rather an uneventful year owing to the remains of my illness, which kept me pretty much a prisoner to the house all the winter & but for the cold spring, we should have gone to Fontainebleau early & returned for May, but the cold kept us till May 10th when we started & were joined by Emma & Lucy, our time at Fontainebleau was spoiled & shortened through Lucy's caprice & early in June we went on to Seeligsberg & remained in Switzerland till August 19th, when we returned to Paris, & were met again by Emma. Frank went to the Brit Association at Bristol & returned to us for ten days. We all went home Sept 14th & prepared to visit the Groves, but were stopped by Lady G.'s illness, so we went to Emma at Leamington & [settled] in Rutland Gate October 11. Emma came soon after. We went to Julian Hill for Xmas but only staid 3 days. Spencer had rather a troublesome year with illnesses. The Russell Gurneys went to Egypt in November. Frank busy about the statistics of his sweet peas. Wrote a paper on Feb 9 in Philosophical Magazine on 'Law of Error & Ogives' & later on 'Twins' & 'Theory of Heredity'.

1876

Another year without great events to us. Frank worked again at the Sweet Peas helped by Miss Christie.[8] We spent Easter at Leamington. Soon after our return I caught cold & was kept to the house nearly a month & last the Geographical Soiree & the opening of the Loan collection at S. Kensington, to which Frank sent several Exhibits & read a paper there about his whistles & their overtones. Emma came to us in June and Darwin's illness at the same time spoiled her visit. In July we visited the Jenkinsons at Ower Cottage[9] and greatly enjoyed the week. In Augst. we went to the Groves at Syston near Bath & met Mrs Hills and Mr Plowden spending a delightful fortnight. We left England on Aug 23, met Emma at Calais & with her & Mr Brodrick travelled through Bavaria & Tyrol & to Venice & the Italian lakes. We returned October 23rd much better, & stronger & had a quiet domestic settlement to my unspeakable comfort. Arthur's engagement to H. Edgeworth

---

[8] Possibly Alice M. Christie the translator, of the 'Kew and Richmond Centre'.

[9] In Fawley, Southampton.

was announced in Aug. Georgina has been many weeks ill & is still an invalid.

1877

A year of illness for me, but marked by so much kindness love & affection, that in looking back, it seems full of sweetness, and to have brought me nearer to Frank & to the dear friends who solaced me, especially to my very dear old & early friend E. Gurney, Amy Lowder too was full of help & in my own family, all were kind & sympathizing & Montagu gave me many kind visits, but Frank was my great support throughout. I only fear that I shall never be able to go about with him as heretofore, and that cuts off so many sources of happiness, but with so many blessings and comforts. I will only look to the bright side of life. My illness lasted from Jan 12 till June, on the 21st we went to Tunbridge Wells, Emma & Temple with us & they were most kind in arranging every thing to save me & help Frank. After this we went to Bournemouth for 6 weeks, and Frank went to the British Association at Plymouth in August & made several short excursions on his return for the day only. Then in September we had dear Emma's kind hospitality for a fortnight & returned home the end of September. Few events to record. The telephone is one of the great discoveries of the year, & Stanley's feat of crossing Africa. Arthur was married at Easter, just when I was at my worst & not expected to live. We have seen so little of him of late years, that his marriage makes little mark in our lives. Xmas has been spent quietly at home. Mr Holland died on the 26th December at Paris.

1878

A year of no great mark, for ourselves and our families. My health has gone on improving, but greatly narrows our life or rather lives, as I lose so many opportunities of making friends & of mixing myself with the stirring interests of the day by refusing all dinner invitations and yet I am so much stronger that I always hope I shall be able to resume old habits, still years go on. Frank had rheumatic gout in his knee last March & was a prisoner nearly 3 weeks, his great solace was working out his ideas about composite photographs, for which he has received great encouragement from men of note, indeed this has been his effort of the year, of course his meteorology has occupied him constantly, but that has come more as regular routine. We again spent Easter with Emma, & returned intending to enjoy our friends & society, but the death of dear Mr. R. Gurney May 31st greatly saddened our summer & Amy Lowder's engagement to Sir L Pelly made another break in our society. The blank caused by Mr. Gurney's death & Dearest Emily's shattered life will never be filled up to me. The summer was neither fine nor warm, with few exceptions. We went to Vichy in July for the waters, & both of us benefited greatly but the after cure at [Mount Dore] was not successful, the weather was wet & I could never make any excursions. F. enjoyed his mountain walks. After we went to Tours & Blois & Paris, but I was not strong enough for such a rush as the exhibition brought so Frank took me to Folkestone & left me for a week, whilst he returned to the Paris Exhibition. After this we went to Leamington for a fortnight & came to Rutland Gate early in October. The winter began early & was severe all through Xmas, and I have been chiefly kept to the house. We have had a few dinner parties but our life is much quieter than it used to be, as I go about so little with Frank.

Arthur's son & heir & Spencer's 6th son are the chief family events of the year. There is a good deal of distress & depression, many failures of Banks. Among our friends, Lady Grove is a great invalid & I fear likely to remain so. M. North has been in India the whole year painting & making friends.

1879

Very little of great moment to us. Gave more dinner parties, but seldom able to go out. Dear Lady Grove died March 8th. We went to Bournemouth for a fortnight in April & after had visits from Lucy & Emma. Went to Vichy in July & after a pleasant 3 weeks met Raes and Martindales, went to Gérardmer which I did not like, as it always rained & the Hotel was crowded. We went to the Black Forest where I had great enjoyment of the lovely scenery & left a bit of my heart behind. Then we paid Emma a visit & Judge Grove at Wakehurst in Sussex & were at home the middle of October to London fogs & a most dismal winter. F. gave a lecture at the R. Institution on Generic Images in May & we paid C. Darwin a short Sunday visit. Arthur's little [Olive] born. Gertrude sent to Bombay in November.

1880

Gave many dinner parties for Frank's sake, but seldom went out to any. Visited Emma at Easter & Claverdon & no sooner settled at home than we were sent off to Vichy late in May & had no summer enjoyment after for we were both ill after & I could neither enjoy our visits nor eat. We went to Boulogne in August & F. returned for the British Association at Swansea. After that, we went to the Meuse & Spa & Brussels Exhibition, nothing very attractive but Spa did me good, then we visited Emma and returned home early in October. Kept very quiet and dull & missed many friends. Montagu's little Maud born in May. I enjoyed meeting Mary and the children at Boulogne, but saw little of Spencer. F. began to photograph at home. Busy arranging the house after its decoration.

1881

A quiet uneventful year & rather dull, for my health puts a stop to so many things of interest & when I am without pain after eating, I am so thankful that the fear of bringing it on makes me fearful of extending my small limit. Frank has been very much taken up by his photography. The winter lasted long, we visited Emma at Easter & then Bessy & Lucy & others visited us & I was overdone, so as any pleasant engagements came, I had to give up nearly all in turn. We went to Stanmore in July, & on to Bournemouth for a month waiting for the British Association Meeting at York. Frank went to Lord Winlock's & we should have met in Devonshire, but constant pain for 6 weeks drove me again to Vichy & we the rest of the season & came home direct from Paris. Dean Stanley died in July & left a deep impression on me & for some time I did hope Montagu would have been his successor. Teddie had a serious illness in the autumn & gave the greatest anxiety but thank God he was restored. We spent Xmas at home. Douglas Galton broke his ankle bone on his way to the south.

1882

Another year & now there seems little to record, though a great deal to be thankful for & a good deal of quiet happiness. Cook bother tormented me till summer & were [sic] a veritable torment, as my dear Frank can little understand, so he was not sympathetic. I enjoyed a visit to Claverdon & Emma at Easter, the weather was pleasant & I love a village church, it reminds me of Gayton. The Moilliets lost their boy in Feby & were in deep sorrow. Frank began his book on Human Faculty early in the year & it had gone on through the year, & it was a great pastime to him during our summer ramble on the Rhine in the Black Forest, Constance & lastly Axenfels. Bad weather haunted us, but we were happy, & I kept well, & began sketching again. It was such a boon not to be kept by a British Association meeting this summer. Mr Darwin's death in May cast a deep gloom over us. M. North's Gallery at Kew was opened in June, & we took Emma & Agnes to it. Soon after, Amy Galton was married at Lambeth. Frank's portrait was taken in June by Professor Graeff & though a great consumer of time, the success was sufficient reward. I just scraped through summer in London without my old pain & by leaving early, but was very worn. Frank spent several days at Guy's Hospital about photographing the patients and seeing what Dr Mahomet had done. On our return from abroad we visited Stanmore and Emma; at the former, I was ill, at the latter Emma & her illness turned to whooping cough, so we were not in luck, coming home we visited Arthur. Cold weather began with our return & kept me in its grasp more or less for 6 weeks, still it was less bad than often & to my infinite joy. A very mild Xmas prevents our realizing the season. George was made Canon of Winchester in June & left Liverpool. The Harrow young ones growing up are a great pleasure. Georgina has been very delicate for months. Extraordinary year for rain & floods. Besides what I have mentioned as Frank's work during the year, he gave a lecture at Eton on Anthropometric Registers & life histories, and wrote a paper in the Fortnightly on the same, & gave one at the Committee of Medical Association.

He was invited to lecture at the Lowell Institution in America but refused. In meteorology he designed a clock for cumulative temperature. He is elected on the council of the R. Society. Was begged to be President of the Anthropological but refused.

1883

The last day but one of the year. It seems such a short one, events for me are fewer and fewer as health impedes my doing much, & this year has been an obstinate one for my pain. Early in Feby dear Georgina's illness took a serious turn & she died after five days severe suffering on the 12th. It was a frightful blow to all of us feeling the immensity of the loss to dear Montagu & his children, besides the love we bore her. That same week Henry Smith died after a very few days illness. Frank attended both funerals on consecutive days. We spent Lent quietly & went to Bournemouth for Easter March 25 to be near the Montagus & Arthur & [Harrie] joined us, it was a pleasant meeting & the weather which was bitter at first became warm & spring-like at the end of the fortnight & gave us a delightful day in the New Forest. On our way home we spent an hour with George at Winchester; Josephine was unwell, so we did not stay the night. In May, Spencer's 7th son was born. We remained in London till July 11th when we went to Stanmore. I was done up by London whirl & grief for Mr. Spottiswoode's death on June 28 & the funeral in the Abbey July 5th & became so unwell at the Jenkinsons that we began our summer outing at Boxcombe & Bournemouth & spent a pleasant month meeting pleasant people, all the time I was on starvation diet. Then we went to Newton Abbott near Torquay & visited Totness & Dartmouth & Torquay, also a pleasant time & with nice day weather such as one seldom enjoys in England, still I prefer a foreign climate & think it suits my tiresome ailment better.

We were kept very anxious about Mrs. Bunbury, Frank went to see her & found her better, she has had many ups and downs since & is now in a very critical state. We went to stay with Emma the middle of September. May Galton was there, it was a constant bustle & Frank got a boil on his back & when we came back in October 1st I was beginning to suffer again from my pain, but we had a quiet settling in and have gone on happily up to now December 30. I have paid several visits to Harrow. Spencer is adding to Julian Hill. Teddie has twice come out head of the school this year & Harcourt is doing well. Frank went to the British Association at Southport in September & in the early part of the year he corrected proofs of his Inquiries into Human Faculty, which was published by Easter. He worked at means of measuring the unconscious helped by Croom Robertson. In the August Fortnightly he wrote an Article on Medical Life Histories, offering prizes up to £500. Spent much time in the details. His Record of Family Faculties just came out, also the Life History Album, which he edited. He was chairman of the Anthropometric Committee of the British Association, which published this year its 4th & final report, also chairman of local Scientific Societies.

Dear Adèle Bunbury died on the 31st December. She hoped she would not live to see the new year & her wish was granted. She was weary with pain & suffering. Thus I have lost two sisters in law this year, both dear to me. But of course dear Georgina's loss is the most grievous for Montagu's sake and the children.

1884

The year seems to have come round with ever increasing speed & I do less and less, owing to uncertain health & constant recurrence of my pain. I seldom go out in society & thus lose the chance of making friends & pleasant acquaintances & more & more home seems the most fitting place for me & for Frank, as he is always full of occupation. This last year he has been chiefly occupied in collating & abstracting from the Family Records, for which he offered prizes to the amount of £500. These records about 150 were sent in by May 15th, and the statistics they afford will be food for many a month & possibly year. Then he took great interest in establishing an Anthropometric Laboratory at the International Health Exhibition at S. Kensington & gave no end of time & money to its prosecution & gained full recognition from Sir J. Paget & others, whose opinion he cared about. He was Rede Lecturer at Cambridge on May 27th & we went to the Vice Chancellor's & after to Mrs Darwin's & quietly enjoyed the 4 days, for fortunately, I was well up at the time having just recovered from a very bad influenza attack, which shut me up for 3 weeks & prevented my going with Agnes to the opening of the R. Academy, on which my heart had been long set. Our absences from home were a week at Brighton in early Feby & a fortnight at Ventnor (Medina Cottage) at Easter. It was bitterly cold returning but Frank & I had some pleasant walks & talks there & enjoyed the holiday & wild flowers. We were there from April 1 to 17. In the summer my great desire to go abroad was stopped by the outbreak of cholera in S. France & after long debate, Frank settled on the Lakes & I was quite willing in default of my pet scheme which I especially had cherished, as there was no English British Association to spoil our holiday, it being at Montreal. So we went to Windermere on July 15th & settled ourselves on the 17th at Keswick & after a week of rain had a good spell of fine weather & enjoyed the walks around, & a pony helped me & friends were very kind to us, still I missed the tonic air & food of foreign lands & my pain began mid Augt & has more or less troubled me ever since. After Keswick we spent a week at Patterdale, a charming place & nice hotel & good food compared with our lodging at Keswick. At the end of Augt we went home for a week. Frank [sighed] after the Health Exhibition then we went to Claverdon for a few days & to Emma for a fortnight. The great heat there finished me up & I was quite bad when we returned home Sept 24th & at once consulted Chepmell.[10] On Oct 7th we went to Stanmore & staid till the 11th little thinking it could be our last visit. It was a nice & happy one, the Alcocks were there & dear Mrs Holland so kind & interested in all we were doing & seemed so well, but alas she took a chill & after a week's illness, congestion of both lungs. She sunk so peacefully on Nov 29th. It is a grievous blotting out of past happiness & forward hopes & a very great loss to me & to many. She is not the only friend we have lost. Sir Bartle Frere died in May & old Mr Bentham in Sept. I saw him only a few days before he died.

Dear Gertrude returned in May from Bombay & lightened the hearts of all her belongings, she improved sufficiently in health to return in Novbr to her Bombay work. We have had several visits from the Spencers, themselves & children, not many from the Galton family; Milly & Baron were prevented coming in June as fixed & this also kept away Emma & Lucy. We have had a quiet home Xmas after a gala on Xmas eve at Julian Hill, where we met dear Montagu & had to congratulate him on Ted's exhibition just gained at Trinity. So the year has ended & I wish I could have done more for those I love, but alas my powers are sadly clipped. Emily Gurney has been away great part of the year. The friendship of the Groves is always a pleasure.

Public interests this year have been greatly exercised about Gordon's mission to Khartoum & the Franchise.

---

[10] Chepmell, Dr. Isaac Dobrée. See page 354.

1885

The early year about Jan 9th Baron Lethbridge died suddenly & all our sympathies were called forth for Millicent & her children. We visited Emma for two days in Feby, as she had been ill & my pain diminished from Feby gave me freedom. At Easter Montagu organized a Family Meeting at Harrow, probably the last, owing to his appointment to the Deanery of Gloucester & we all met in the Chapel on Easter Sunday. Besides this Frank & I spent a week at Tunbridge Wells & a little more at Ramsgate. Then we settled at home & found M. North returned from Chile & had several pleasant meetings with her & finally went together to Holmbury St Mary the end of July for 6 weeks & thoroughly enjoyed the fine country & pleasant drives.

Early in Septbr Frank presided over the Anthropological Socy at Aberdeen at the British Association meeting & I went to Emma, who was kindness itself. Owing to cholera in Italy, we gave up our intended visit there & after a dull week at St Leonards & pleasant visits at Oxford we settled at home. Each year, it is like a fresh beginning of life & difficulties meet one but with all these, home attracts & with special force when absent from it. I have often had longings to travel abroad again & feel quite rusty with long English residence. Servants and F's occupations are the impedimenta often & sometimes my health & having to take Chumley.[11] Now I hope for it is 1886, if all is well. The great family event of the year has been Montagu's removal to Gloucester, it may not make much difference to us but I feel to have lost the children. The Elections have been a great autumn interest. Frank as President of the Anthropological Socty has enjoyed their meetings & worked hard to get good people. Also he has been instrumental in getting up a Geographical Exhibition of maps & globes & methods for improving English geographical education under the superintendence of Mr [Keltie], in rooms at Great Marlborough Str. The Bar of London (Temple) has this year appointed Arthur Preacher at Whitechapel.

1886

The year is gone & left us in cold & snow as it began or rather severe cold set in the 6th Jany & lasted with little change till

[11] The maid.

Easter; but we left England on Feb 25th for Paris & escaped it & had a delightful time at Hyères first, then Mentore, Rome & Sorrento met George & the Meuricoffres. We were so happy there, every passing had its tale of keen enjoyment & we felt to grudge each day as it passed. We returned by Amalfi Castellamare & Florence where we spent another week to be remembered such beauty & loveliness in art & nature. We hurried on to reach Vichy for Easter, having promised a visit to Milly there, it was burning hot & I was unwell, but we enjoyed dear Milly & her nice children & the glorious mountains & lake. We came to London May 5th & soon had a visit from Montagu who was in trouble about Edith, lung disease having declared itself & we were deeply distressed by the death of Imogen Hall & the grief of the dear Groves. After a month of home & seeing various family friends, I had a fall on my temple & never got round the rest of our time in London. Emma paid us a short visit in July & then we went to Contreseville for Frank to take the waters & get rid of gouty symptoms. Mrs. McLennan went with us & we all grumbled at the dullness of the place & were glad to escape at the end of 3 weeks to Bussang in the Vosges Mts, a really nice place, beautiful woods & excellent Hotel. We came home in great heat by Nancy, Paris & Boulogne. We were too demoralized by the heat to like to try another time, also M. North was anxious to have us. We went to her at Mt. House Alderly Spt 1st & had an amusing visit of a fortnight. Lotta Symonds was there in part, from there we visited the Hills at Newberry & then to poor Montagu at Gloucester. Edith had been ordered to Davos & Agnes had married Mr. Howson Sept 14th, but Ted & [Hugh] were with him & Emily came. It was a time of gloom, but we were glad we went, for our sympathies were very keen. Thence we went to Claverdon & Leamington & home early in October to receive Emily & Montagu. They left on (before) the 8th for Davos, a thick fog & lurid atmosphere the day of their leaving & many forebodings crossed our minds; 3 weeks after, we had the joyful news that Montagu had accepted the Mastership of Trinity. George has had a serious illness he caught fever in Naples & in July he was struck down with rheumatic fever & another attack in the autumn. Dear Arthur too has had anxiety about his home party. The Spencers have taken a house in London & left Julian Hill. I can truthfully say, I have felt stronger this past year, the terrible pain has barely troubled me & Frank has been well & active. Early in the year, he sent 2 memoirs to Roy Socty on Family Likeness in stature & in eye colour, shewing law of each parent contributing $\frac{1}{4}$ each, grand parent $\frac{1}{16}$th etc. He presided at Anthropological conferences at the Colonial & Indian Exhibition. He gave a lecture at Birmingham on Heredity, & at Toynbee Hall the same week in Novbr & received from the Royal Society a gold medal for his biological researches. Politics have been in a most troubled state but I trust Gladstone is gone forever & that Lord Salisbury will help us through. Each year seems shorter than the last & this one more especially from the long break of our Spring tour.

1887

I have less to record than usually, for there have been few events to mark our home details, and happily no great trouble. The winter lasted severe till nearly Whitsuntide & ended by 2 days of pouring rain while we were at Tunbridge Wells for a short change. We spent 10 days with Montagu at Trinity in early March & our drains were thoroughly redone meanwhile. At Easter Montagu came back joyful from Davos about his dear Edith, but she soon drooped & returned in May to England to see the new home but to die on June 20th. Jubilee Day was the 21st, I was glad not to hear of her death till we returned from the beautiful sight of the Queen's Procession to the Abbey from the Athenaeum, a sight to be remembered. About July 16th we went to Homburg for Frank to take the waters & staid 3 weeks, we enjoyed the fresh air, the music out of doors the pleasant sitting under cover but were glad to leave. Thence to Freiburg Constance & Zurich, from there Frank made a 6 days tour to the Rhone glacier, Eggishorn, Rieder & Belalp & Brieg to Domo d' Ossola. San Rosso & the Falls of Toce & back to Zurich over Gries to Rhone Glacier. Then we went to Engelberg for a week & by the St Gotthard to Locarno & Lugano. After 2 days of bliss there, Frank fell ill with a sharp gastric feverish attack. At the end of 9 days, we travelled to Basle & the next night direct to Calais & home. After a few days rest, we went to Emma, but F. fell ill with Bronchitis & again we had to hurry home & soon after he had an attack of the Coecum (Typhlitis) all in 2 months. This last interfered with the 3 lectures he had prepared to give at S. Kensington Museum on behalf of the Anthropological Institute on Heredity & Nurture, but a fortnight later he was able to give the 1st. He wrote an article in the July Fortnightly on "Good & Bad Temper", also a memoir for the Entomological society on Pedigree Moths Breeding which Mr. Merrifield undertook & has made good progress in writing a new book "Natural Inheritance". So end our chief records, few enough, to make the year appear shorter than ever, & to make me most gratified for the small amount of troubles, which mark it. My health too has been very fair, except the local pain in my face for some few months, which always however culminated most inopportunely.

1888

This has seemed a very short year one sign of advancing years, because one has fewer events to record from the inability to take part in as many as heretofore. Frank was ill all the early part, congestion of the lung & I felt very anxious about him & overtired myself, as when he recovered I fell ill & that brought on my old pain, which never left me till we had recourse to Vichy in Sptbr. By this I was prevented visiting Montagu at Trinity & lost many nice things & felt myself an encumbrance, unable to make plans with friends. We spent a week with Emma at Easter & a very cold one, indeed a very [seedy] year & a very wet summer. Montagu took us by surprise in June by the announcement of his engagement to Agnata Ramsay the Senior classic of 1887 & the wish to be at his wedding modified our summer plans.

We staid in England visiting the Lethbridges at Clifton & then Marianne North, at the end of a fortnight, we returned to London for Montagu's wedding Augst. 9th. It was a very interesting gathering & picturesque arrangement at St Margaret's Westminster. Next day we went to Brighton for a week that Frank might see & watch his pedigree moths. The air was very keen & affected my liver so I did not enjoy the change & went back gladly to Mount House.[12] But it rained & it rained & though we were very happy & many nice visitors, the Wallaces, Kegan Pauls, Mrs Ross, Proctors, my pain got worse & the damp increased it & Frank persuaded me to go to Vichy, we went Sept 1st & had 3 weeks of glorious weather, a flood of sunshine on our rooms from early morning, it just gave me fresh life & we basked in the dry warmth & were happy.

[12] M. North's residence.

Frank was busy about his book, "Natural Inheritance" correcting proofs & I led a quiet life, as suited the cure, taking the waters & we did so enjoy the fresh morning air. The late season prevented our meeting pleasant people & the last week Frank was a sufferer from a bad fall on his right arm & this hurried us home after 3 days in Paris, instead of making a stay at Campagne, as I hoped, so we were back the end of Septr & soon comfortably settled & home rest was a good conclusion to the Vichy cure.

We paid visits to Arthur, Mr Brodrick, the John Hollonds (a farewell to beautiful Stanmore) & to Trinity College to make an acquaintance with Agnata. We greatly enjoyed it & her & Queenie & the Grimthorpes came the last day for the great college commemoration & we ladies listened to the speeches in Hall. We staid Xmas week with Emma & came back in a dense fog. Now for Frank's special work which I write from dictation. Frank started Anthropometric Laboratory at S. Kensington, built the place himself on the site given rent free. About 1000 persons measured during the year from March. Much interested in Bertillon's method of identifying prisoners by measurements. Gave Friday evening lecture at R. Institution on Personal Description & Identification, memoir to R. Society on correlations. At Anthropological, memoir on head measurements of Cambridge students & one on Australasian Class Marriages. Continued to be President. Worked fully during the year at "Natural Inheritance". Pedigree Moths still going on at Brighton but data not yet ready to work out.

Few family anxieties this year, though the last 2 months George has been ill with Jaundice & our dear M. North has been most seriously ill & made us very sad, but I hope she is mending. It has been a very sunless year & the absence of any summer weather cast a gloom over those bright enjoyable months. Aunt Rebecca died in Novbr in her 91st year, from Bronchitis. She was ill three weeks. Politically the Emperor Frederick's death in June has left marked consequences, following so soon his old father, just 3 months. I have been busy & happy the last 3 months & free from my old pain to my great joy & thankfulness.

1889

Our first Sunday of 1890, I will sum up the last year, I always shrink from doing it but I must make myself do what I have made into a duty, though probably only Frank will ever cast his eyes over our mutual record of happy days.

We have both been very fairly free of illness. We went abroad in early March till May 3rd & very likely our warming in the south & open air life steeled us against illness. Mrs. Maclennan went with us & was an excellent & very patient travelling companion, but essentially dull & matter of fact, I always felt oppressed by her & yet bounden by her complaisance. We went to Cannes to see George, then to Mentone, Genoa, Milan, Baveno Cadenabbia & home by the St. Gotthard to Paris where we saw the Eiffel Tower & exhibition buildings & hurried home to escape the Babel opening. When we returned it scarcely seemed worthwhile to settle in to London & I feel the Spring tour put our year a little awry. We remained till July 24 but lost our dear Emma's visit, as she came ill & left in a few days & I was ill with a cold those few days. We spent nine days with M. North & were pleased by feeling we gave her pleasure, she was cheery but suffered a great deal from the dropsy. Then we returned to London & Frank left next day for Paris for a congress on heredity, of which he was chairman & gave an address published with some mistakes of meaning in Revue Scientifique. He had sent Anthropometric Instruments there at the opening of the exhibition, He returned in a week & on Aug 12th we went to Carlisle & spent 10 days with the Hills at Corby Castle & a fortnight after at Whetheral (Miss Little [Leaf] Dine) We made many pleasant excursions in the neighbourhood full of interest & beauty to Naworth, Lanercost Hesham, but I never really saw the Roman Wall, we were prevented staying at Chollesford. We were with E. Wheler at Alnwick for 3 days, but hurried to Newcastle on Sept 10th & were very kindly entertained till the 18th by the Spence Watsons at Bensham Grove Gateshead. I liked them very much but could not really enjoy the B. Association meeting. We were too distant from the sections. Frank read a paper about marks in examinations for physical efficiency & another about an apparatus with pendulum for reaction time. We saw Durham Cathedral & York & so on to Leamington for Lucy Wheler's marriage Sept 26, the interest of which was meeting Milly. We came home early in October & late in Nov. spent a few days at Trinity Lodge. Montagu was alas ill & the baby James also. Frank has been busy all year on problems connected with correlation, but has published nothing yet about them. He took up thumb marks to illustrate the difference between genera & species & was perpetually working at them during our spring tour. His book on Natural Inheritance came out in Feby but has elicited but small notice, it is beyond the range of most minds. Bettany has republished in the Minerva Library his travels in S. Africa, a very attractive volume. He succeeded Mr De La Rue as chairman of the Kew committee. His Laboratory at S. Kensington under Sergeant Randall has gone on steadily. The year has been dull as to weather, very little sun, a great deal of darkness & if I have a right to long for any thing, it is for more light & for the power of doing more & being less taken up by small duties. I am always rejoicing in the absence of my "old pain" everything seems light compared with that. Poor old George is the one anxiety in our family. Emily has bought a house at Oxford. Our dear England threatened by Democratic Strikes.

1890

I hope I am not ungrateful for the many blessings we have enjoyed this year, but as it has just passed away, I feel as if it were the duller and least interesting of any I remember. The two first months, Frank had so much cold & bronchitis affection that he we had to decline all evening engagements & now, since our autumn settlement I have been laid by with a very sharp attack of my old gastric enemy, so our social gatherings have been nought. In March dear George's death saddened us all & came very unexpectedly, as we had been kept in ignorance of his real state. Soon after we went to Alderley to spend a fortnight, including Easter, with our dear M. North & though grieved at finding her far less well than the previous summer, I recall those days with great pleasure & interest. The weather was fine & how we basked & enjoyed the sheltered wood walks & how I loved the early spring flowers. Poor Marianne I did hope to see her again, but she became much worse in June & I felt loath to leave England. Still we hoped for another rally & went to Switzerland in July; not a successful journey, though we liked both Beatenberg & Stoos[13] our chief resting places, but we met few English, no friends, & it was dull & the places we should have liked did not suit Mrs. McLennan & I always felt her a drag & a

[13] Stoss.

hindrance with all her kindness. The weather was generally wet & I suffered very much from the chill damp & discomfort at Freudenstadt, food bad & wind every where. Returning we came by Strasbourg & Nancy & Paris & there the sad news awaited us of dear Marianne's death on Aug 30th. It was such a saddening blow we had no heart to seek enjoyment so we came to England September 8th & soon went to Leamington for a visit & came home the end of the month. I hoped to do a great deal in the weeks before Xmas, but as I began to sicken on October 7th, all such hopes have been fruitless. Gertrude returned from Bombay in Sept very worn & much aged. I have not seen her since September. October was a beautiful month as to weather, but since, a most severe winter has set in & promises to last some time. Stanley's return from Africa in early summer was the excitement of the season & then his marriage with Dorothy Tennant. Frank has been very well ever since Easter & must give me his notes of work done to note down. The Groves have launched two boys, Harcourt to India, Walter to the Navy. Owing to Parnell's evil doings, politics are more hopeful, as he Parnell has forced a split in his party. After our Easter visit to Alderley we spent 2 cold days in the Forest of Dean, which Frank enjoyed & one delightful walk I had with him on the high wooded banks above the Wye where we had the offer of a dead badger. During our visits Frank was always busy on Thumb marks & wrote & read an important paper on them for the Royal Society in November which is being printed in their Transactions. He was continually being hindered in his work by objections made by Dixon[14] to the 1st step in his theory of correlation & this entailed much correspondence & fruitlessly, at last Mr. Watson took his view & will write a memoir thereon. Frank published a paper in Nature by which dice could be marked & used for experimental work in probability questions. He made a machine (the account published in the Anthropological Journal) for measuring the velocity of movement. He spent a good deal of time in calculations as to the effect of natural selection & on the advantage of sexual reproduction, but is not satisfied with either. The laboratory has been at work during the year.

His pedigree moths are in the 8th generation, bred by Mr. Merrifield & others. Miss Andersson has measured 7 generations & Miss Reynolds has photographed them. Frank read a

[14] Dickson, James Douglas Hamilton. See page 356.

paper at the Society of Arts in November on Physical Tests in Competitive Examinations & has reason to believe that a system of applying them has been decided between the India Office & the Civil Service Commission. This entailed no end of labour and correspondence. Frank served in a committee of the Royal Society on "color vision" under the sanction of the Board of Trade. At Kew they determined to test photographic lenses. In the early part of the year, he acted as chairman of the Meteorological Council during General Strachey's absence in India & among other things had the Weather Boards placed in the balcony. Thus he has been busy the whole year on his abstruse statistics which consume time & thought & cannot be expected to have immediate recognition. Dear Emma was not strong enough to pay us her usual summer visit, alas for our high stairs. In the Galton family E. Wheler's marriage has given great satisfaction. So ends this year & may the next give a better retrospect.

1891

Jan. 10. 92.

Our first Sunday of this year at home & very cold & gloomy, snow everywhere, though not deep. The year has been wet & sunless & few bright sunny days rise up before me, but we have not had illnesses as many & I trust may withstand the present cold. The illnesses I had in late 90 kept me to the house till about Feby, when I began to emerge, but I never felt equal to evening engagements & gave up my yearly Gala, the ladies soiree at the Royal Society, which seemed like lopping off another branch of pleasure. We spent 10 days with Emma at Easter. She was full of kindness, but too full of feasting & I got cold & gave trouble to my regret & felt I had no right to pay visits till stronger. In May we spent 2 days at Cambridge with the George Darwins & I enjoyed that & seeing the dear ones at the lodge & a drive with Montagu. It is a great pleasure to feel capable of what formerly came about naturally ie. visits & friends; of more general society, I longer aspire to, I feel it beyond me, but I love my friends & am truly blessed in many; I only wish I saw more of the Arthurs and Montagus, but their special work divides us.

Our summer was cut up through Frank being President of the Demographic Section of the Congress of Hygiene held in London from Aug 10 to the 15th, so we went to Tunbridge

Wells the middle of July for about a fortnight, returning in time for the Congress & as soon as it was over, we started for Vichy, where all went well as to health, & very fair weather, but as we had a dull time, no friends to help out the days. We left Sept 9th for Chatêl Guyon, & after Royat, on which we fixed to afford a meeting with the Pellys, this took place on the 22nd & then we made our little excursion to the Lozère, the Gorges of the Tarn, Frank's long ambition, Montpellier-le-vieux Nimes and Avignon, a very pleasant fortnight together, though I paid for it, by taking a chill & had inflamed eyelids for some weeks after our return on the 9th of October. It seemed a pity to have to turn Northwards as the days were becoming cold, but we could not well forsake home & servants & I felt even then to need more warmth & comfort than we secured moving from place to place. Since then till the end of the year we have been at home & dined with the Alcocks on Xmas day, feeling it very likely it might be their last, both are so delicate.

In Frank's work, he was exercised in Spring by a sudden demand to vacate his laboratory, but a new building on the grounds of the Imperial Institute was given him & the usual work carried on. Dr. Garson was temporarily engaged to help.[15] Frank's chief occupation has been on finger marks & in the tabulating he has been largely assisted by Mr. Collins of Birmingham & Miss Jeffrey. He published a way of indexing the patterns in the R. Society's Proceedings. He also has made enquiry into Racial peculiarities by finger prints at the Jew's schools, from Wales & of Negroes from the W. Coast of Africa. Mr. Merrifield continues breeding moths, but they are not sufficiently numerous for statistical purposes. He is still Chairman of the Kew Committee & has helped to arrange methods for testing photographic lenses sketched by Capt Abney,[16] but worked out by Leonard Darwin. He has gone on with the Colour Blind Committee. Quite lately he has joined a committee for the Advance of the Knowledge of the Hausa Language a pet idea of Dr. Barth. This is a full resume of his year's work.

Montagu has ceased to be Vice Chancellor & not too soon, as an attack of Influenza has touched his heart & for a while he is ordered to give up writing & extra engagements. So has passed our year, too happy in many ways, & too blessed in all

[15] John George Garson (1854–1932).

[16] Sir William de Wiveleslie Abney (1843–1920).

essentials to think of disappointments, which also lead one to feel our life closing in & the dim great future looming ever & ever nearer.

1892

Jan 1. 93.

Alas the year has flown & how little I have done for others & myself, but illnesses & the fear of making myself ill by imprudence cut short my resources. We spent the 1st week of the year with Emma & a few days after our return, Frank fell ill with Bronchitis, happily I could nurse him & did not catch any influenza, which was extremely prevalent & fatal. He was in his room about a fortnight, but out of doors at the end of 3 weeks, though obliged to take great care & avoid cold. This made him anxious to go South & we went to Biarritz the end of March. We came in for a good deal of bad weather at first & I took cold, which culminated in a real illness at the end of a fortnight, when we went to Cambo & I suffered terrible pain in my head quite overwhelming me by day. In the midst of this, I heard of Sir Lewis Pelly's sudden death at Falmouth where he had gone for the Easter holiday. How this has saddened & affected my life I cannot say, he was a true friend, but my heart has sickened for dear Amy's immense loss & prostration & how I miss their genial pleasant intercourse, it is a fearful break for the poor Alcocks also. We returned to England early in May, but did not see my dear Amy for some weeks. Our London summer was very short & not eventful, it took me some time to recover strength. In early August (8th) we made another start, first to Corby & enjoyed 10 days. There was a succession of pleasant people, Andrew Langs, Pollocks, etc. & the Hills most kind & hospitable. Thence to Edinburgh & took rooms 28 Walker street & staid over a fortnight. The 1st week was very enjoyable, the beauty & associations of the Queenly City delighted us & the excursions to Dalmeny Park, the Forth bridge, Dalkeith, Roslyn[17] etc., but at the end of the week, our weather broke & we had scarcely a fine day after, nor at Callander where we went next, but we did have a good day for the Trosachs & Loch Katrine. I got no sketching & we both felt the cold & damp & went away without regret on the 19th September to

[17] Roslin.

the E Wheler's at Alnwick. We quite enjoyed our 3 days visit, but not the cold, how it did pierce in the drive to interesting Warkworth castle. On the 23rd we went to Lady Welby at Denton Manor near Grantham, with a beautiful house & garden & Lady W. so attractive and mystical & taking up with her writings & Frank. We left on the 27th & spent a little time at Peterborough on our way to Leamington, a chill autumnal day, wind wailing, dead leaves falling, decay all around of what had been. Even the Deanery had a sad look, Dean Argles lay on his sofa, quite changed & enfeebled & we felt there was a good man passing away, as he died in a few weeks.[18] Well we reached Leamington in the evening & had a warm welcome, but somehow the rain & cold still prevailed & I feel as if we have entirely missed a summer this year, I hope it will not always be so. We came home early in October to receive Milly & her Amy & then Catherine Symonds came,[19] but here again I was ill from cold & acute pain in my left side, so I lost much of her visit. She was a delightful visitor & if I had been well, I should have truly enjoyed her. She came like a legacy from dear M. North & was full of interest. This illness also prevented our going to Montagu & we have had a very quiet Xmas at home & since the 23rd severe cold. Frank's chief business of the year has been connected with bringing out his book on "Finger Prints", in which he was much helped by Mr. F. H. Collins, revising the MS. & also he worked at the statistics, but without very good results. In connection with this, Frank has now set up Photography at his laboratory in order to make a series of standard specimens. He has also brought out a new edition of his Hereditary Genius. His only paper published this year was a short one in the American Science on Boston Schoolboys, pointing out a close relation in stature between the variability & mean at different ages & that the probable error & the mean value may rarely be as independent as is commonly surmised. He attended an Experimental Psychology Congress in August, of which he had declined to be President, he did not go to the British Association at Edinburgh but was there appointed Chairman of a committee to report on an Ethnographic Survey of the British Isles, which is now going on on a practicable scale. As Chairman of Kew Committee he took part in the establishment of

[18] Marsham Argles (1814–1892), briefly Dean of Peterborough.

[19] Janet Catherine North (1837–1913), sister of Marianne North and wife of John Addington Symonds.

Major Darwin's photographic lens testing. Still on the Geographical council & is one of the Hausa Committee which was started by missionaries & took part in the election of a student. Still V. Chairman of the Meteorological Council &, in General Strachey's absence, has served as Chairman. He was asked to send certain instruments to the instrument exhibition at Nuremberg including the Quincunx & did so. We are well & happy in each other & in our home & good servants, but the year has gone rather uneventfully though kindly. Tennyson's death in October has left a great blank.

1893

December 31.

The last Sunday of the year & a true winter day, as we turned into the park after church, there was thick hoar frost on the grimy green grass. How quickly Xmas returns, swifter each year, as our events lessen, but we can look back on a glorious time of sunshine & feel its rays still in memory, when we think of its blessed influence at Hyères at Valescure at Cannes & finally through our English summer. We both had influenza in Feby & went Riviera March 13th taking Mary Chepmell with us; all was pleasant, but above all the unexpected meeting with Montagu & Agnata at Valescure. We returned home April 20th & remained till May 11, when we visited Montagu at Trinity & staid till the 15th. Cambridge in full beauty of Spring freshness & very enjoyable. In July we were alarmed about Millicent & she came to London for an operation under Mr. Treves,[20] which seems to have cured her, time will show.

As soon as she was out of danger we left for our holiday at Holyhead & spent a pleasant 3 weeks at "The Institute", Miss Adeane[21] helped us in all ways & was always planning for our good, shewing us the chief interests of the Island in delightful scrambling drives, thence a short visit to Lady Stanley at Penrhos, after which we settled ourselves at Bettws-y-Coed, a lovely place, but Frank had an asthmatic cough & that prevented our full enjoyment. After 3 weeks there we sent Chumley home & made a delightful round on coaches to Capel Curig, Beddgelert & Aberglaslyn to Port Madoc, a heavenly crisp day & then by toy Railway to [Ffestiniog]. The next morning on through Bettws-y-coed to Conway & Chester. I

[20] Sir Frederick Treves (1853–1923).

[21] Of the Holyhead Institute and Training Home, for girls in service.

was rather disappointed there, because I had expected a great deal from the picturesque old houses, but the Cathedral was very striking & so were the old battlemented walls. We spent 2 nights at the Hotel & then on to Shrewsbury to Leamington, we had only 2 hours there, but drove about & saw the chief points & then wandered about the castle walls, till the time to go. We staid with dear Emma a fortnight & came home Sept 30. I was anxious to return early as Frank declared gravely he should not spend the winter in England, but so far he is home satisfied & is very well, his deafness has certainly increased & that makes me less socially inclined, though feeling isolated in consequence. My nephews George & Cyril have married this year & Montagu has another boy born Dec 20. I have been stronger, the effect I believe of the warm sunny year. The Symonds are in London for this winter, which is a joy, the girls are full of spirit & intelligent usefulness.

Frank gave a lecture at the Royal Institution on Jan 27 on the "Just Perceptible Difference" which included a method of telegraphing outline drawings. The rest of the year he has mainly given up to finger prints & Identification questions. He made many photo enlargements of them for standards. Through corresponding with Lt. Col Surgeon Hendley of Jeypore an order was passed for all Indian army recruits to be finger printed. He got repetitions of prints at Hooghly of 8 men, whose imprints had been taken by Sir W. Herschel 14 years previously & wrote a small vol. on them "Decipherment of Blurred Fingerprints". He drafted a letter from Council of British association to make enquiry on identification & wrote to Times thereon. A committee from the Home Office visited his Laboratory several times & will report. The year has been most weariful in politics, Gladstone & Home Rule ever to the front & a bar to all great patriotic endeavours. We have lost a friend lately & a true one to a large circle, Professor Tyndall. I look back on his lectures & former balmy days as surrounded by a halo. As I write the year is nearly gone & I feel it has been a good friend to us & merits our gratitude. May God bless and keep us during the one coming & help us to lead lives useful & helpful in our narrowing circle lest I "rust unburnished not to shine in use".[22]

[22] Tennyson, Ulysses.

1894

Jan 13. 95.

Alas I have let the new year 95 run on till now before making a resume of 94, idleness & real suffering from severe cold which began on the 27th have made me delay. The prominent points in the year were our Spring journey to Villa Garin Cimiez & to the Tyrol & Bavaria in late Summer. We left home for Cimiez Feb 27th & returned home April 7th, Mrs McLennan went with us & became quite cheerful & was very helpful till her fall near Vence & the Saut de Loup. The Woodroffes made Cimiez very homelike to us & we both loved & enjoyed the place, though weather was less favourable than the previous Spring. But then we went earlier. After our return the usual home life went on diversified by endless troubles about young servant girls, who give me a very low estimate of the MABYS institutions for girls.[23] We went in May to Edmond Hills at Darland House & on Jun 1 to Trinity, & went on the 2nd with Agnata to an afternoon "at home" at Newnham. A very interesting meeting of students & their friends & we saw some of the girls rooms. On June [14] we went to Arthur at Oxford for Frank to receive a D. C. L. degree, having had a pleasant visit to Merton May 12-14 to meet Amy Pelly quietly, 4 visits in 2 months were quite eventful for our quiet routine. The summer was very uncertain as to weather & put a stop to country expeditions. Emma did not come till July 12 for her visit & only staid a week & then I began to be quite done, a hot week early in July when Arthur & Montagu were here gave the first blow, so we had to begin our holiday by drinking iron water at Spa for a fortnight, after which Mrs McLennan joined us at Coln & we travelled on by Heidelberg to Boden See thence by Arlberg Railway to Landeck Innsbruck & Berchtesgaden, where we staid nearly a fortnight & enjoyed it spite of bad weather. After this to Salzburg, which was a real treat & on by the Lakes to Ischl, of these Hallstadt & [Altaussee] were the most striking, but for one & all, for proper appreciation one ought to spend some time & we were fully late, it turned so cold. We emerged at Gmunden Sept 17, & spent 2 happy days there & then by Regensburg Nuremberg & Würzburg back to Coln, these cities I should have thoroughly enjoyed, but for a long sharp attack of diarrhoea so I saw nothing but the outsides & returned home

[23] Metropolitan Association for Befriending Young Servants.

rather disconsolate on Sept 27, feeling the great opportunity wasted so far as I was concerned. After a few days at home to refit we went to Emma for a week. Darwin began then to be uneasy about Fanny & she died on Nov. 9. We had a pleasant time till after Xmas, when cold set in & we have both suffered from it & the past sunny days seem to melt from memory.

In Frank's work, the Home Office Committee on Identification made its report & was accepted by Government, so all prisoners are hence-forth to be finger printed. Experiments are going on in Bengal to introduce the full system there. Frank took considerable pains in revising his system of classification going repeatedly through his large collection of Finger Prints & writing a book on it (not yet in the printer's hands). He spent many weeks over a problem about risks of error in classification. He wrote an article in Mind on discontinuous evolution & one in the National Review on the part of religion in evolution. He was appointed chairman of the Committee of the Royal Society for Measuring Plants and Animals, Prof. Weldon doing the real work. He has given a park fence to the Kew Observatory, on condition of receiving the additional ground needed to protect the Observatory from Golf Clubs etc. With my family happily all has gone well & the young ones are pushing on in different lines. Harcourt Butler was married in August in India & brought his bride over for a short time. We met all the Spencers in England at their Xmas gathering & I felt what life & vigour there was among the young ones: ambitions & hopes. With us a quiet yet happy decline, blessed in our home when so many around us of intimates are bereft of their dearest. I often feel a dread of what may come to either of us after 41 years of happy union, but let us trust our heavenly Father that he will help us at all times, as he has hitherto helped us.

1895

Always it costs an effort to record the year till I am ashamed of my idle procrastination & now I am determined to begin the task. The beginning of the year 95 began with great cold & we had a trying time. Frank fell ill Jan 10 & kept half invalided till the 21st when Dr Chepmell[24] came & advised rest and quiet. He attended the Meteorological Council on Feb 6 &

[24] Chepmell, Dr. Isaac Dobrée. See page 354.

became worse on the 9th, with gastric catarrh Dr C called it, any way, he was very miserable & thought himself very bad & for a few days I was very uneasy about him as he got weaker & made no progress, so he went one better & worse up to March 7th when the weather & he both improved & he began to go out & about in a careful way & was able to attend a Kew Committee on the 29th. We intended a change to Hastings, it was crowded by Influenza Patients, so we went to dear Emma on April 9 for 9 days & there to Mrs. McLennan at Hayes for a short visit. Towards the end of the month weather became warmer. Emily came for the 1st week in May. Mrs McLennan on the 8th to 10th little did we think, it was her last visit. Cambridge gave Frank a D. Sc. Degree & Montagu gave us his delightful hospitality for the event May 13 to 16, two of the days were perfection of Spring cowslips, buttercups, blue sky etc., then a blast like ice chilled the air for a few days, till May 21 when real summer came & lasted with little intermission till the end of September. Another pleasant country visit was to Mrs Hodgson at Tanhust at the bottom of Leith Hill, the Rhododendrons in fullest bloom & trees in fullest foliage. This was our last taste of country, Emma came to us on the 7th of June for a week & Frank became impatient to leave London, so we missed several of our summer friends & festivities & left England on July 3 for Nauheim to see Arthur & Emily, who were taking a course of Baths & thence to Garmisch in the Bavarian Highlands to Villa Baden, a Pension kept by Mrs Bethell, she made us very comfortable & herself very pleasant & we staid more than a month, the scenery was very fine, but there was a drawback, i.e. so little shade & the weather was very hot with frequent clattering storms, still I enjoyed it & so did Frank. We made a mistake accepting Teddies kind invitation to Corby for September, having travelled so far, I would have been better to remain in the distant lovely country, but we left Garmisch on August 12th for Munich, caught cold there & had no heart for Munich sights, save the Pinacotheca, the city looked so badly & widely laid out, great need of organization to unite fine units of buildings, any way it was discordant after our rural life at Garmisch, we thence travelled to Nuremberg, which we both loved, though Frank's ideal of the previous year had somewhat waned. We were nearly baked there & nearly burnt out at Rothenburg our next place, both were full of old world mediaeval life, with churches & fountains & Rath Haus & picturesque bits of buildings.

Nuremberg is modernized to great extent & rather hides the old, but the two exist & excite one's wonder & admiration. Only being able to get food uneatable for hardness, we gave but 2 nights to Rothenburg, though full of interest & sketches to attract, but we frizzled & journeyed on to Frankfort & Bonn & home on Aug 27. I soon fell sick from the great heat & a horrid crossing & we had to give up Corby to our sorrow & spent instead a prosaic fortnight at Tunbridge Wells, again very hot, but we had nice drives in the fair English country, Hop gathering going on everywhere, & went to Emma on Sept 21 till the 30th & then home to a clean arranged house, soon to be disarranged for the installation of the electric light, which has been a great boon to us through the dark winter days. Our good & devoted friend Mrs. McLennan died on December 10 after 3 months of great & increasing suffering from a tumour on her hip probably the effect of her fall at Vence by Cimiez; we shall feel her loss to the end of life. Another friend has passed away, Mrs Jenkinson, & no tidings of her illness have reached us & our gay young friend Mary Chepmell has married & gone to a home in Burmah so our little circle narrows alas & I have not the energy left to widen it much, for my digestion has been very uncanny for 4 months & makes me fear evening society. All the Butler family have kept well except George, he has had to resign his post as C. S. Examiner[25] & he & Cyril are gone to Algeria to recruit & Mia spent Xmas with us & we 3 dined at the Spencers on Xmas day & enjoyed the gleeful capacities of the children. The year has ended sadly for England threatened by United States, Venezuela, the Ashantis & war rumours from the Transvaal. Well we here have a strong Government under Lord Salisbury & favourable auguries for public working since the summer elections.

With Frank the year has been unusually barren, partly the effect of influenza & in some degree to increasing deafness, which makes attendance at public meetings useless & at committees of little avail. He published Finger Print Directories in May, & thus completed all he wished to say on the subject. Shortly after he closed his laboratory & gave the instruments to Oxford in charge of Prof. Thompson. He is glad to find his work on heredity & correlation beginning to elicit the interest of mathematicians, notably of Karl Pearson. Lately he has been busy on methods for procuring measurements of pedigree stock, especially by photography with corrections for effects of

[25] George Grey Butler (1852–1935), Civil Service Examiner.

perspective, but is not quite prepared to publish the principle yet. Besides the Kew Committee & Meteorological Council, he is chairman of Royal Society Committee for Measuring Plants and Animals. Gertrude left England this day for India, nominally for 6 months. Spencer's sons have gained many laurels this year. Indian C. S. Navy & Woolwich & Arthur's Harold a brilliant success at New College Oxford. Mrs. Wheler has been on the sick list for 5 months but is now getting on. So ends the year, I feel towards it as a real friend; having learned to withdraw from many pleasures & attractions, I find great happiness in the quiet enjoyment of home life, in books in home duties & friends & in the promise of success & useful distinction of the young members of the family & in the hope of keeping their love. One great regret is the barrier Frank's increasing deafness occasions in our intercourse & of this, he has little idea; it also makes society difficult, so that we are not fit to mix in any large circle. Gertrude is gone back to India for 6 months & we are glad, she can escape a rigorous winter, dear Emily Gurney has been sorely crippled by sickness & illness & would be glad to pass away to her last rest. I hope for better things here, both she & dear Amy Pelly are a constant source of watchful love, thank God for granting them & may he bless us all in love & peace during the New Year.

1896

This is the 1st Sunday of 1897 & I must make my notes of the past year, which has run its course calmly for us & our families but has grievously impoverished us by the loss of close & dear friends & their blanks can never fill. Dear Emily Gurney died October 17 after a mysterious illness of 7 months, it began with lung inflammation soon after the opening of her chapel for "the Weary" & from that time she was dead to all her friends, it did seem hard to know she was lying seriously ill & to be kept away without a word or a smile as always greeted us; our old friend Sir W. Grove died Aug 1st our wedding day, we lose not only him, but the Harley Street house & Mrs Hills & many intertwining interests. In the early year, Frank had a good deal of cold & cough & thought himself very bad, so we went to Hastings, March 13, for a few days & to Eastbourne on March 25 for a week & to Emma on April 2 for Easter. Soon after our return Darwin announced his engagement to Miss Cumberland, which gave me a pang to think how soon a dear life companion can be forgotten for the sake of comfort & companionship. We had our usual visitors, in May & June, Milly on June 18 & Harrie on the 22nd, most unfortunately, Frank fell ill on the 19th with gastric catarrh like he had before, but much milder & it might have been nothing, had he not gone to Kew with temperature 102 so with my guests & nursing I was nearly done for. Chepmell advised Wildbad as a cure & though still very weak but in perfect summer weather, he & I left England July 10 & began the treatment July 14 & also the rains began & marred our holiday & I may say they have continued ever since with but little variation, so Frank's after cure at Oberstdorf did not prosper & we were glad to escape to Lindau to see blue waters & blue sky again & feel warm. I do love more & more light & sunshine & open air life. From there we went to Davos & enjoyed the Symonds & their beautiful home & our Hotel Boul. For 10 days, then snow came & melting snow was not salutary & our patience under contre temps waned & we took the Symonds advice & turned homewards by Zurich. We had no heart to establish ourselves short of Italy & there also floods & tempests, so we went for a Cathedral tour to Freiburg, Strasburg & Rheims, which more than exceeded our expectations & to Amiens, but dripping rain curtailed our visit there & we had a happy return home Sept 7th. Remained till the 25th, then a short visit to Emma & back to settle for winter. Oct 2 Montagu & Gertrude paid us visits during the Sept sojourn & Gertrude joined the R. Catholic church, where her heart had long been. We went to Trinity Oct 13 for a few days & were happy, dear children, but Montagu terribly cast down by Archbishop Benson's sudden death & he had to leave for the funeral at Canterbury. Then we settled at home for winter & have gone on quietly since, happily so far very little cold & we have both thriven on home life. Frank is especially well, but he complains his brain power is not as vigorous as formerly & he cannot work quickly. Late in autumn, he again took up "Pedigree Stock" & induced the Royal Society Committee of which he is chairman to undertake "Darwinian" work & to utilize existing means for experiments on heredity, what they may lead to can be judged later on. The Kew Observatory thrives, Douglas Galton has worked hard to get it merged into a National Physical Laboratory & an influential deputation will shortly wait on Lord Salisbury about it. Frank wrote a short paper in the October Fortnightly upon Telegraphic Signals with Planet Mars.

He is always occupied but has less to show for it this year than often, partly owing to Bronchial colds. We have alas lost our good friend Chepmell, he has resigned & left London for good, we shall miss him sadly, in Frank's illnesses. I felt his support firm & satisfying, indeed after 22 years of his wise advice, I cannot imagine how we can live without him & his kind wife. So surely do good things come to us & pass from us, & I try to be thankful for the innumerable blessings we have had even with the pain of feeling them gone. So ends our year, not an eventful one, but a calmly happy one ending with a merry Xmas at Spencer's, the young felt full of life & ambitions, Monty left in November for India & is now at his post at Hissar in the Punjaub.

1897

It is with painful reluctance that I set down the incidents of this fatal year, and do so on Jan 6, the anniversary of the day when I first became acquainted with dear Louisa at the Dean's house, next door to our own, at Dover in 1853.

In the early part of the year I was more of an invalid than she was, but we had some pleasant outings together, as to Nansen's great meeting on Feb. 8.[26] Chiefly on account of my persistent asthmatic cough, we went to Bournemouth, March 24, partly to be near Dr Chepmell, whose remaining eye was threatened. He told me to go to Cauterets or Royat. Montagu and Agnata came to us for a day, from Lyndhurst, while we were there. We had had alarming news from time to time of Emma, from middle of Feby onwards. At length she was better & we went to her April 20-23. Louisa was well enough for some small festivities, a tea party her last, on May 7th & the military tournament. We went to Oxford, to Arthur's, June 5-8. June 21 Jubilee days we went with Mrs Lyell to the Athenaeum they had excellent places & Louisa was not over tired. Next day Bessy came to tea & Mr Lyell. 26th I went to Naval Review, L. was not strong enough to go with me.

July 14th left to Royat, slept at Boulogne; next day, a weary waiting till 10pm at Lyons Station but night journey comfortable, Louisa not suffering at all. July 24th Puy de Dome with W. Livett & a young lady. L. remained in the garden at the auberge while we went up & had luncheon set

[26] The Norwegian Dr. F. Nansen, who was honoured in the Royal Albert Hall for his exploits at the north Pole.

out. I never saw her more pleased & nicer as a hostess than when we came down. Aug 1 Mme de Falbe arrived, in far from good health.[27] Aug 3, L. awoke with diarrhoea—we all had it—but recovered. Very sultry. Arranged for Pont St Laurent in Dauphine and wrote to have letters sent to Grenoble. On Sunday 8th she was apparently quite well and half packed for a start next morning. Monday 9, she had had bad diarrhoea & sickness in night, not worse than frequently before. Tuesday Aug 10 was worse, I had Dr Petit in, who made light of it, but said he wd come the next morning. Wed 11th she was very ill, but saw Mlle de Falbe who was able to leave her bed for the purpose. L. wrote a post card to prepare Chumley in case she was wanted. In afternoon very weak indeed continued vomiting. Thursday 12 worse & in a very serious state. That night, or rather on Friday morning early at $2\frac{3}{4}$, she quietly passed away. On Saturday Aug 14 she was buried in the cemetery of Clermont-Ferrand in plot 419, which I purchased as a concession in perpetuité. So our long married life came to an end. Writing as I do now, after nearly 5 months have passed, and am better able to take a fair retrospect, I think that the inevitable blow occurred at a more seasonable period than at any other time. Dear Louisa's vigour was distinctly declining; she was still able to enjoy much, but was I fear rapidly on the way towards permanent invalidism, and she was conscious of a weakening of her mental powers, small things fatiguing her much more than formerly. Had I died first, I fear her strength would have been inadequate to carrying on life unaided. She has been in many respects a most valued as well as a loved example to me. May her good influence abide, though she personally is gone. All her friends lay stress on her power of sympathy.

How I pulled through the terrible strain and hurried requirements of the occasion, I cannot conceive, but I did, thanks largely to the hearty and tactful sympathy received from M. de Falbe & W. Jennings, who had made Louisa's acquaintance & returned to help. I could not leave Royat on account of letters, till Tuesday night, arriving in London Wedy afternoon where the sympathy of Spencer & Mary & of Gertrude awaited me. Some few days were spent in sorting her clothes and in carrying out Louisa's wishes. Then to dear Emma at Leamington for a week; thence to the Douglas Galtons at Himbleton also for nearly a week; thence to Mrs Hills at Corby; all of which

[27] Falbe, Eleanor Louisa. See page 356.

greatly braced me. The general kindness of her (Louisa's) & my relatives was extreme. On returning Sep 13, Frank Butler was ready to live with me; a most valuable help against the sense of isolation. Sep 25 we went to Boscombe, saw Milly & went to Weymouth to see Emma, who was there. Oct 16 to G. Brodrick for 3 days Oct 23 to Emma at Leamington. Dec 4 to Bessy first, then to the Studdys.

The family events of the year are,—Emma's serious illness & complicated ailment. Darwin's slight but repeated paralysis. Mrs Leonard Darwin's very serious illness. Sir Rutherford Alcock's death. Isabel Butler's marriage.

My own occupations were the inquiries into the Bassett hounds, which led to the "Average Contribution of Each Ancestor &c." Proc. R. Society — also inquiries into speed of American Trotters Proc. R. S. & note Method of Photo measurement of horses &c., published today Jan 7/98. The committee on a Phys. National Laboratory has, & is, taking evidence. Evolution cttee has not done much; Kew Obs prospers; Meteor. Council, usual routine.

Register

Afrikaner, Jonker

c1780–1861. The leader of the Oerlam Afrikaners between 1823 and 1861, the second oldest son of Jager Afrikaner. Though his exact date of birth is unknown, he was reported to be on commando in the early 1790s, so he must have been born by roughly 1880 at the latest. The Nama nicknamed him 'wart-suckle'. When he succeeded Jager, a split developed in the group, with one faction led by his elder brother Dawid, and the other by Jonker, who moved further north with his followers. Jonker resumed the destructive but lucrative raiding that his father Jager had forsworn in exchange for a pardon. He targeted the southward-pressing, cattle-rich and militarily weaker Damara (Herero), initially as the 'protector' of the Nama Red Nation. An astute leader and military commander, he greatly expanded the scope and influence of the Afrikaners. 375–377

Afrikaners

A clan of Oerlam Hottentots, outlawed in the Cape Colony and resident in South-West Africa in the region of Eikhams, near Windhoek, at the time of Galton's expedition. They were headed at that time by Jonker Afrikaner, who had succeeded his father Jager Afrikaner. Before that Jonker's grandfather Klaas Afrikaner had been in charge of the clan. 383

Bushmen

Sometimes referred to as 'San' or 'Saen'. The diminutive but oldest inhabitants of the Southern African region, with a language containing distinctive click sounds. By the time white settlers arrived in the area, the Bushmen had already been marginalized by the incursions of the Bantu to semi-arid regions, where they foraged. The Hottentots and their sub-grouping the Nama were close relatives. 382

Chepmell, Dr. Isaac Dobrée

(1828–1919), educated at Elizabeth College and Kings College, and for at least 8 years the travelling personal physician of Lord Hollond. He was forced to retire in 1897, due to acute glaucoma. His other patients included Lady Gregory. See *Lancet*, Feb. 6, 1897, 394 349, 367

Concordiaville

The name given by the Wesleyan missionaries to Eikhams, seat of Jonker Afrikaner. In Jonker's time this was just outside Windhoek to the East, and was thus Klein Windhoek or 'Little Windhoek'. A hot spring was located nearby. The place had other names too, including Elberfeld, "Queen Adelaide's Bath", Esek, /Ai-//gams ('High Water'), and Otjomuise. 378

Cookes, Denham Charles Johnstone

1810–1847.iv.21. Educated at Worcester College Oxford (BA, 1834). He was the son of Denham James Joseph Cookes, of Worcester. Galton first met him in London circa June 10, 1843, shortly before the tour he made to Germany with Emma Galton, Harry Hallam and Harry's sister Julia. 'We then returned to Portman Square to await the appointed call of Denham Cookes. He has the funniest head I ever saw, is exceedingly agreeable, and at his ease; nobody except his lawyer knows where he lives, under cover to whom all communications are addressed. His hair is yellowish red; his face something like this

He told us a great deal' (Pearson 1914, 179). He later shared a carriage with Galton between Paris and Avignon, as Galton set off on his Middle Eastern trip of 1846–6. They stayed together for a few days at Avignon. 'He was charming and full of anecdote and fun' (Galton 1908b, 199). In April 1847 Cookes was killed when he was thrown from a horse at a steeplechase in Florence. 'It is our painful duty to record the death of Denham Cookes, esq., which took place under very melancholy circumstances, at Florence, on Wednesday, the 21st ult. On the preceding Monday the deceased was riding in a hurdle race a horse of his friend Lord Ashtown's, when, in leaping one of the hurdles, the horse fell; Mr. Cookes was thrown on his head, the horse falling heavily on him; he was taken up senseless, and notwithstanding every care and attention he never rallied, but expired on the night of the 21st ult., very greatly beloved by a numerous circle of friends, who will long and deeply deplore his untimely end' (*Devizes and Wiltshire Gazette*, Thursday 13 May 1847). 219

Cornelius

c.1800–1867. A Nama Chieftain, also known as //Oaseb. Leader of the 'Red people,' or Khai//Khaun, the longest-settled of the Namaqua groups in South-Western Africa. He was based at Hatsamas on the upper Skaap River, south-east of Eikhams (the closest modern town is Dordabis). He was, like his Oerlam rival Jonker Afrikaner, an inveterate cattle raider. 24

Crackanthorpe, Montague

1832–1913. Son of Christopher Cookson of Wellington. Educated at St John's College, Oxford, taking a double first in classics and mathematics and winning a fellowship. Practiced successfully at the bar, Q.C. in 1875. Took the name Crackanthorpe after inheriting the Newbiggin estate in Westmoreland in 1888. Stood as a Liberal for Parliament twice, becoming a unionist. Lived at 65 Rutland Gate, and so was a neighbour of Galton, who lived at number 42. Wrote widely in the periodical press, covering subjects like population control, which led to a book in 1907, *Population and Progress*. On June 25th, 1908, he hosted a meeting of the Eugenics Education Society at his house where Galton spoke on eugenics, which led to widely increased membership when reported. Succeeded Sir James Crichton Browne as president of the society, at Galton's desire.

261

Dickson, James Douglas Hamilton

1849–1931. Scottish Mathematician, at Peterhouse in Cambridge. He collaborated with Galton on the early stages of the mathematical theory of correlation (specifically, correlated jointly distributed normal variables). 358

Dixon, Peter Daniel

c.1820–? A trader, partner of James Morris and Thomas Lawton of Cape Town. Born in Cape Town. Active in the Walfisch Bay area since late 1843, trading from a post in Sandfontein. 20

Eikhams

See Concordiaville. 375–377, 384

Falbe, Eleanor Louisa

1822–1899. Born Eleanor Louisa 'Lucy' Hawkes. Married to the Danish Ambassador, Christian de Falbe, her third husband, who had died in 1896. Previously married to John Gerard Leigh (1821–1875) of Luton Hoo. Prior to that, to Humble Dudley Ward (1821–1870). 373

Frere, Sir Bartle Edward

1815–1884. Governor of Bombay, later British High Commissioner to Southern Africa, 1877–1880. In 1872 he was sent on a mission to the Sultan of Zanzibar, returning in 1873. 129

Galton, Darwin

1814–1903. The oldest of the brothers Galton, a country gentleman. He served in the militia, and took part with the Warwickshire Yeomanry in the quelling of chartist riots in Birmingham in 1839. He was married three times: to Mary Elizabeth Phillips (1822–1869), in 1840; to Fanny Jane Arkwright (1827–1894) in 1873; and to Penelope Maria Elizabeth Cumberland (1832–1912) in 1896. The Edstone estate that he at first farmed, near Stratford-upon-Avon, was acquired through his first wife,

Mary Phillips. His son John Samuel Phillips was born in Nov. 1841 but died less than a year later in August 1842. He served as High Sheriff of Warwickshire. Later he moved to Claverdon Leys. 'My elder brother Darwin was a great favourite among his friends from his early life onwards. He used me as his fag when I was a boy, and taught me to be fairly smart. I imbibed many common-sense maxims from him, but our ideals of life differed to an almost absurd degree: he had not the slightest care for literature or science, and I had no taste for country pursuits. Our differences of temperament became more marked the older we grew'. He was noticeably eccentric and had a deep aversion to music. In mid-life he had a series of epileptic fits. "He was an old-fashioned conservative in politics, and had a profound reverence for Church and State. Indeed he often stationed himself near the G. W. R. [Great Western Railroad] to salute the Royal train as it passed, and it is said even at night." (*Warwick and Warwickshire*, Jan 10 1903).

'DEATH OF MR. DARWIN GALTON. [*Leamington Spa Courier*, Friday 09 January 1903, 8.] We regret to have to record the death Mr. Darwin Galton, which occurred at his residence, Claverdon Leys, Sunday night. Mr. Galton had reached the age 88, and had been for some time past in failing health. For many years Mr. Darwin Galton occupied a prominent position in Warwickshire. He was Deputy-Lieutenant, and, with the exception of Lord Norton, the oldest Magistrate in the County, his appointment dating back to 1840. As long ago as 1850 he was High Sheriff for the County. He also for many years commanded a troop in the Warwickshire Yeomanry. When the County Councils were formed in 1888, he was elected Alderman, a position he resigned some three years ago. Mr. Galton was born March 18th, 1814, at Ladywood, near Birmingham. He married, in 1840, Mary Elizabeth, eldest daughter and co-heiress of the late Mr. John Philips, of Edstone. The only issue of the marriage was a son born in 1841, who died in the following year. He married, secondly, in 1873, Frances Jane, third daughter of Mr. Peter Arkwright, of Willersley, in the County of Derby. She died in 1894, and Mr. Galton married, thirdly, Miss Penelope Cumberland, who survives him. The family of Galton have been long connected with Birmingham. Mr. Darwin Galton's great grandfather, Samuel settled there about the middle the 18th century. He lived at Duddeston House, then a country residence, but now in the heart of the town. Mr. Samuel Galton was a large manufacturer of small arms for the Government, was a member of the Society of Friends, and died in 1799. His son, John Samuel, and his grandson, Samuel Tertius, father of Mr. Darwin Galton, for many years carried on a bank in Steelhouse Lane, known as Galton's and James' Bank, till it was finally closed in 1831, and the following year Mr. S. T. Galton came to live in Leamington, where he resided till his death in 1844. The connection of the family with Leamington has never ceased, and two of his daughters, Mrs. Edward Wheler, and Miss Galton [Emma Sophia] are still living here. Mr. Samuel Tertius Galton bought a small property and built a house at Claverdon Leys. This passed to Mr. Darwin Galton. who has lived there since 1858, and has added materially both to the house and to the estate. He was Lord of the Manor of Claverdon and Henley-in-Arden. Besides the two sisters above named, two brothers survive him, Mr. Erasmus Galton of Loxton, in the County of Somerset, and the well-known scientist, Mr. Francis Galton, whose work as a meteorologist, and anthropologist, has world-wide reputation. Mr. Darwin Galton was a first cousin of Charles Darwin, the eminent naturalist, both being grandsons of Erasmus Darwin, M.D., F.R.S., of Derby, the author of "Zoonomia" and many other works. Mr. Galton is buried to-day in the family vault at Wootton Wawen churchyard. The funeral will be very quiet in accordance with his expressed wish. Mr. Darwin Galton was, throughout his long life, a strong Conservative. He took a prominent part in the old division of South Warwickshire, and when the County was divided under the Redistribution Bill of 1885, he greatly assisted in the organisation of the Party in the South-West or Stratford-on-Avon Division the County. At the formation of Polling District Committees he was appointed Chairman of the Claverdon polling district, a position he occupied until the time of his death. He presided over all public meetings in his district, and his speeches commanded attention, for he always sought to persuade those who differed from him. In purse and person he was always willing to help the cause he had much at heart. He presided at the last public meeting of the Arden Habitation of the Primrose League in August last, and was present at the annual tea and garden party of the habitation, held at Ardencote, the residence of his sister-in-law, Miss Phillips, when Sir James Sawyer spoke of him as "the natural head of the Conservative Party in this division of the County."'. 337

Galton, Frances Anne Violetta

1783.iv.23–1874.ii.12. The wife of Samuel Tertius Galton. Mother of Francis Galton. Daughter of Dr. Erasmus Darwin, the successful physician and natural philosopher, by his second marriage to Elizabeth Chandos Pole. 337

Galton, Louisa

1822–1897. Born Louisa Jane Butler. Daughter of George Butler (1774–1853) headmaster of Harrow and later Dean of Peterborough, and Sara Maria Gray. Wife of Francis Galton, whom she married in 1853. Died at Royat in France of a gastrointestinal complaint. 337

Grove, Justice William Robert

1811–1896. Active in the Royal Society, which he was responsible for reforming on meritocratic lines. An expert on patent law, he was educated at Oxford, and had early interests in chemistry and electricity. Galton knew him from the Royal Society, and spent a number of vacations shooting with him. Grove contributed a piece on 'Nabloos and the Samaritans' to Galton's series *Vacation Tourists and Notes of Travel in 1861*, Galton 1862d. Galton was also friendly with his daughter, Mrs. Anna Hills. 338

Hottentots

Relatives of the Bushmen, with some Negro admixture, and a pastoral economy often supplemented by foraging. Frequently referred to as 'Khoi' but no name is settled. The Nama were a sub-group of the Hottentots, whose Cape branch rapidly acquired Europeanized culture and admixture, as in the case of the Oerlams. 376, 382, 383

Nama

A sub-division of the Hottentots, also known as the Namaqua. They were located north of the Cape, pushing up through Namaqualand into South-West Africa. They had some Negro admixture but were most closely related to the Bushmen, or San. 24, 376, 377, 382

North, Frederick

1800–1869. MP for Hastings, where he lived at the Lodge. He was the father of Marianne North, a friend of Louisa Galton. He was a widower after 1855. The Galtons often stayed with the Norths at Hastings. 337

Oerlams

Acculturated Hottentots with some European admixture, many of whom spoke the dialect of low Dutch that they had acquired from their masters the Boers. The Afrikaners were a prominent clan of Oerlams. 375, 382, 384

Tyndall, John

1820–1893. An Anglo-Irish experimental physicist from County Carlow, based at the Royal Institution in London. He was educated at the University of Marburg in Germany. Tyndall was a frequent walking companion of Galton—they were both members of the Alpine Club. They may have first met in the Alps in 1857 when Tyndall was climbing the Jardin in the company of T.H. Huxley: 'We met Mr. Galton, the African traveller, returning from an attempt upon the Jardin; and learning that his guides had lost their way in the fog, we deemed it prudent to return' (John Tyndall *The glaciers of the Alps, & Mountaineering in 1861* (London, 1861) [Everyman, 1906]: 59). Tyndall's emphasis on visual demonstrations of physical phenomena, in his lectures at the Royal Institution, may have influenced Galton's use of analogue demonstrations of statistical concepts like the Quincunx. Galton is known to have attended those lectures. 'My late but passionate love for mountaineering was one cause that subsequently brought me into frequent contact with Professor Tyndall ... who was then at his very best physically and mentally. He, I, and Vaughan Hawkins ... an eminent classic in his Harrow and Cambridge days and of first rank in mountaineering, made a tour together in Cornwall. We chose our way on Tyndall's principle, that it is easy to find difficult places to climb elsewhere than in the high mountains. Certainly he was skilful at discovering them. One of his freaks sent my heart into my mouth. It was at a gully, strewn deeply with loose stones that led over a sea cliff. Down he dashed, the stones were all set in motion like an avalanche, but somehow he extricated himself in time and got clear to one side of them. At another place an isolated needle or cone of rock was separated from the shore by a narrow strait through which the sea swirled, but which could be leapt at low water. We leapt it, and clambered up, he declaring that it was as difficult a bit of rock-work as he had ever been on. We reached the top and got back successfully, jump and all, to the mainland, where I was glad to feel in safety. The Irish dash in Tyndall's blood gave a charm to all he did' (Galton 1908b, 192–192). Tyndall and Galton were allies in the debate over the role of religion in science—see their contributions to the 'statistical efficacy of prayer' controversy in 1871. (Various pieces to do with this, including some by Tyndall and Galton, were collected by the Congregational Publishing Society in *The Prayer Gauge Debate* (Boston, 1876).) A lifelong struggle with insomnia led to Tyndall's death from a self-administered chloral overdose in 1893. 337

Windhoek

Another name for the vicinity of Eikhams. It may have been named so in memory of the 'Groot Winterhoek' near Cape Town where the Oerlams came from, or just because it was, and is, windy there. 376

List of Figures

Bibliography

Anonymous (1810). "Theodore Galton." In: *Gentleman's Magazine* 80.2 (1810), pp. 587–588.

Darwin, Charles (1871). "Pangenesis." In: *Nature* 3 (Apr. 27, 1871), pp. 502–503.

Galton, Francis (1852). "Recent expedition into the interior of South-Western Africa." In: *Journal of the Royal Geographical Society* 22 (Apr. 1852), pp. 140–63.

— (1853). *The Narrative of an Explorer in Tropical South Africa*. First edition. London: John Murray, 1853.

— (1856a). "Jamaica: Interesting Discovery." In: *Notes and Queries* (May 24, 1856), p. 410.

— (1856b). "Signals Available to Men who are Adrift on Wrecks at Sea." In: *Times* (Apr. 4, 1856), 10e.

— (1857a). "Negroes and the Slave Trade." In: *Times* (Dec. 26, 1857), p. 10d.

— (1857b). "The exploration of arid countries." In: *Proceedings of the Royal Geographical Society* 2 (Dec. 24, 1857), pp. 60–77.

— (1858). *Catalogue of Models Illustrative of the Arts of Camp Life*. London: T. Brettell, 1858.

— (1861a). "African Exploration [Remarks]." In: *Proceedings of the Royal Geographical Society* 5 (May 27, 1861), pp. 208–215.

— (1861b). "Du Chaillu's Explorations [Remarks]." In: *Proceedings of the Royal Geographical Society* 5 (Feb. 25, 1861), pp. 110–111.

— (1861c). "On a new principle for the protection of riflemen." In: *Journal of the Royal United Services Institute* 4 (Apr. 30, 1861), pp. 393–396.

— (1861d). "Zanzibar." In: *The Mission Field* 6 (June 1, 1861), pp. 121–130.

Galton, Francis (1862a). "On Baron von der Decken's East African Expedition [Remarks]." In: *Proceedings of the Royal Geographical Society* (6 Jan. 27, 1862), p. 51.

— (1862b). "Recent discoveries in Australia." In: *Cornhill Magazine* (5 Mar. 1862). Unsigned, pp. 354–364.

— (1862c). "Report on African explorations." In: *Proceedings of the Royal Geographical Society* (6 May 26, 1862), pp. 175–178.

— ed. (1862d). *Vacation Tourists and Notes of Travel in 1861.* London: Macmillan, Apr. 26, 1862.

— (1871a). "Experiments in pangenesis, by breeding from rabbits of a pure variety, into whose circulation blood taken from other varieties had previously been largely transfused." In: *Proceedings of the Royal Society* 19.126 (Mar. 30, 1871), pp. 393–410.

— (1871b). "Gregariousness in Cattle and Men." In: *Macmillan's Magazine* 23.126 (1871), pp. 353–7.

— (1872a). "On blood-relationship." In: *Proceedings of the Royal Society* 20 (June 13, 1872), pp. 394–402.

— (1872b). "Statistical Inquiries into the Efficacy of Prayer." In: *Fortnightly Review* 12 (Aug. 1872), pp. 125–35.

— (1872c). *The Art of Travel; or, Shifts and Contrivances Available in Wild Countries.* Fifth edition. London: John Murray, 1872.

— (1872d). "The Efficacy of Prayer." In: *Spectator* 24 (Aug. 24, 1872), p. 1073.

— (1873a). "Africa for the Chinese." In: *Times* (June 5, 1873).

— (1873b). "Hereditary Improvement." In: *Fraser's Magazine* 7 (Jan. 1873), pp. 116–30.

— (1874a). "Appendix C: Statistics." In: *Notes and Queries on Anthropology. For the Use of Travellers and Residents in Uncivilized Lands.* Ed. by John Beddoe. First edition. London: Edward Stanford, 1874.

— (1874b). "On men of science, their nature and their nurture." In: *Proceedings of the Royal Institution* 7 (Feb. 27, 1874), pp. 227–236.

— (1875). "The history of twins, as a criterion of the relative powers of nature and nurture." In: *Journal of the Anthropological Institute* 5 (Nov. 23, 1875), pp. 391–406.

— (1883). *Inquiries into Human Faculty and its Development.* London: Macmillan, Apr. 24, 1883.

— (1888). "Remarks on Replies by Teachers to Questions Respecting Mental Fatigue." In: *Journal of the Anthropological Institute* 18 (Apr. 10, 1888), pp. 157–68.

— (1901). "Souvenirs d'Egypte." In: *Bulletin de la Société Khédiviale de Géographie du Caire* V (1901), pp. 375–380.

— (1903). "Our national physique–prospects of the British race–are we degenerating?" In: *Daily Chronicle* (July 29, 1903).

— (1907). "Vox Populi." In: *Nature* 75 (Mar. 7, 1907), pp. 450–1.

— (1908a). "Address on Eugenics." In: *Westminster Gazette* 31.4730 (June 26, 1908), pp. 1–2.

— (1908b). *Memories of My Life.* London: Methuen, 1908.

Guest, Montague (1903). *The Royal Yacht Squadron. Memorials of Its Members.* London: John Murray, 1903.

Guy, William A. (1859). "On the Duration of Life as Affected by the Pursuits of Literature, Science, and Art: with a Summary View of the Duration of Life among the Upper and Middle Classes of Society." In: *Journal of the Statistical Society of London* 22.3 (Sept. 1859), pp. 337–361.

Pearson, Karl (1914). *The Life, Letters and Labours of Francis Galton. Volume I: Birth 1822 to Marriage 1853.* Cambridge: Cambridge University Press, 1914.

Pückler-Muskau, Hermann (1845). *Egypt Under Mehemet Ali.* Trans. by H. Evans Lloyd. 2 volumes, from the German edition of 1844. London: Colburn, 1845.

Quetelet, Adolphe (1849). *Letters addressed to H.R.H. the Grand Duke of Saxe Coburg and Gotha, on the theory of probabilities, as applied to the moral and political sciences.* Trans. by Olinthus Gregory Downes. First English edition, from Lettres à S.A.R. le duc régnant de Saxe-Coburg et Gotha, sur la théorie des probabilités, appliquée aux sciences morales et politiques, 1848. London: Charles and Edwin Layton, 1849.

Speke, John Hanning (1860). "Captain Speke's Adventures in Somali Land." In: *Blackwood's Edinburgh Magazine* LXXXVII (May 1860), pp. 561–580.

Wallace, Alfred Russel (1870). "Hereditary Genius." In: *Nature* 1 (Mar. 17, 1870), pp. 501–503.

Made in United States
Troutdale, OR
11/03/2026

56845408R00209